Modern Power Systems Engineering

This *Modern Power Systems Engineering: Analysis, Stability, and Control* book bridges the gap between theoretical knowledge and practical application, making it an indispensable resource for engineers, researchers, and transmission and distribution professionals alike. The book equips readers with practical knowledge and industry-ready skills for solving real-world problems in the field of power system analysis, including:

- Designing power system controllers, covering aspects such as network stability, short circuits, harmonic reduction, and more
- Addressing industry challenges such as network instability, harmonic losses, unreliable capacitor placement, wind power plant issues, and low economic production with confidence
- Gaining insights into network issues and how to develop effective solutions with ease
- Applying real-world case studies from diverse power grids in the US, UK, Malaysia, and Iran

The book provides the practical knowledge and industry-ready skills needed to excel in power system analysis and is aimed towards power system engineers as well as undergraduate and graduate students.

Modern Power Systems Engineering
Analysis, Stability, and Control

Mostafa Eidiani and Kumars Rouzbehi

CRC Press
Taylor & Francis Group
Boca Raton London New York

CRC Press is an imprint of the
Taylor & Francis Group, an **informa** business

Designed cover image: Mostafa Eidiani

MATLAB® and Simulink® are trademarks of The MathWorks, Inc. and are used with permission. The MathWorks does not warrant the accuracy of the text or exercises in this book. This book's use or discussion of MATLAB® or Simulink® software or related products does not constitute endorsement or sponsorship by The MathWorks of a particular pedagogical approach or particular use of the MATLAB® and Simulink® software.

First edition published 2026
by CRC Press
2385 NW Executive Center Drive, Suite 320, Boca Raton FL 33431

and by CRC Press
4 Park Square, Milton Park, Abingdon, Oxon, OX14 4RN

CRC Press is an imprint of Taylor & Francis Group, LLC

© 2026 Mostafa Eidiani and Kumars Rouzbehi

ISBN: 9781032968032 (hbk)
ISBN: 9781032968049 (pbk)
ISBN: 9781003590514 (ebk)

DOI: 10.1201/9781003590514

Typeset in Palatino
by codeMantra

Contents

Preface

Traditional textbooks often focus heavily on complex mathematical derivations and university-specific software, leaving students unprepared for the realities of industry practice. This book bridges that gap. The complexity of power system analysis often necessitates specialized industrial software such as DIgSILENT, ETAP, or PSSE. Unfortunately, many recent graduates often find themselves needing to learn these programs through additional and often expensive extracurricular courses.

This book bridges the gap between academic knowledge and industry practice. It caters to a wide range of readers, including undergraduates, postgraduates, recent graduates, and practicing engineers (both new and experienced). Readers can design any kind of power system controller a professional engineer could require, including the best possible network design for both static and dynamic stability, short-circuit protection, harmonic reduction, and more. Professional readers deal with issues such as network instability, harmonic loss issues and capacitor placement, unreliability, wind power plant issues, low economic production, etc. on a daily basis. This book serves as a concise manual for resolving these issues. Expert readers of the power system are aware that sophisticated engineering software is required for real power network analysis. These programs provide a broad overview of the actual network, highlight the network's actual issues, and enable practical solutions. The theoretical material is made simple to understand by these programs.

This book fills the gap between theoretical knowledge and practical application and is a valuable resource for undergraduate and graduate students, power system engineers, and researchers and transmission and distribution professionals. Overall, this book empowers students and professionals with the practical knowledge and industry-ready skills needed to excel in the field of power system analysis.

Since it would be cumbersome to analyze every power network issue in a single book, we have divided them into two volumes. The table below displays data for volumes 1 and 2. We believe that by studying these two volumes, readers will gain insight into nearly all power network issues.

In the end, we had a great time in the writing phase of this book, and we truly hope you enjoy reading it as much as we enjoyed creating it!

Volume 1
Modern Power Systems Engineering: Analysis, Stability, and Control
An Overview of Industrial Power System Software
Modeling a Power Grid

Load Flow Control Methods
Short Circuit Analysis Techniques
Nonlinear Dynamic Analysis Methods
Control Strategies for Power System Stability
Asynchronous Motors Analysis
Power System Harmonic Analysis
Optimal Capacitor Placement Methods
Asymmetric Distribution Networks
Optimal Power Flow Analysis
Reliability and Contingency Analysis
Wind Farm Construction
Large Network Simulations

Volume 2
Modern Power Systems Engineering: Planning, Control and Operational Optimization
Quasi-Dynamic Simulation
Cable Reinforcement and Cable Sizing
Tie Open Point Optimization
Voltage Profile Optimization
Optimal Remote Control Switch Placement
Optimal Equipment Placement of storage models & voltage regulators
Transfer Capacity Analysis Determining
Protection Coordination Assistant
Unit Commitment
Network Reduction
Arc-Flash Analysis
Hydro Power Plant Analysis
HVDC Technology for Offshore Wind Farm Grid Integration
Railway Systems Analysis
State Estimation Analysis
Probabilistic Analysis
Small Signal Stability Analysis

Authors

Mostafa Eidiani (Senior Member, IEEE) earned his B.S., M.Eng. and Ph.D. degrees in Electrical Engineering from Ferdowsi and Azad University, Iran in 1995, 1997, and 2004, respectively. His research interests include renewable energy integration, power system control, transient and voltage stability, power system simulation, and DIgSILENT PowerFactory simulations and analysis. He has authored or co-authored 11 technical books, 9 chapter books, 40 journal papers, and 110 technical conference proceedings.

He is the author of three books published by the Taylor & Francis Group: https://www.amazon.com/author/eidiani

- *Fundamentals of Power Systems Analysis I: Problems and Solutions*
- *Advanced Topics in Power Systems Analysis: Problems, Methods, and Solutions*
- *Fundamentals of Power System Transformers Modeling, Analysis, and Operation*

He was listed among the world's top 2% scientists in 2024, as published by Stanford University.

Kumars Rouzbehi (Senior Member, IEEE) received his Ph.D. in Electric Energy Systems from the Technical University of Catalonia (UPC), Barcelona, Spain, in 2016. From 2002 to 2011, he was an academic staff member at the Islamic Azad University (IAU), Iran. From 2017 to 2018, he was associate professor at Loyola Andalucía University, Seville, Spain. In 2019, he joined the Department of System Engineering and Automatic Control at the University of Seville, Spain. He is the patent holder for AC grid synchronization of voltage source power converters and has contributed to over 120 technical publications, including books, book chapters, journal papers, and technical conference proceedings. He has been a Technical Program Committee (TPC) member of the International Conference on Electronics, Control, and Power Engineering (IEEE ECCP) since 2014 and a scientific board member of the IEA International Conference on Engineering and Management since 2015. He was also a TPC member of COMPEL 2020.

Professor Rouzbehi is an associate editor of the *IEEE Systems Journal, IET Generation, Transmission and Distribution, IET Renewable Power Generation, High Voltage (IET)*, and *IET Systems Integration*.

He is a co-author of the following books:

- *Active Filter Design* (in Persian)
- *Advanced Topics in Power Systems Analysis: Problems, Methods, and Solutions*
- *Fundamentals of Power System Transformers: Modeling, Analysis, and Operation*

1

Overview of Power System Studies and Software Applications

1.1 Introduction

This chapter provides an analysis of various popular power system software applications. Each software has its strengths and specialized uses, tailored to different power system analysis and simulation needs. The overview below summarizes the main features and ideal applications for each software based on the authors' expertise and preferences.

1.2 Power System Studies Software

Several software packages specialize in analyzing and simulating power systems, with the "best" choice determined by specific requirements and priorities. Some of the most popular options are:

1.2.1 DIgSILENT PowerFactory

Strengths: Comprehensive toolset for various studies (load flow, short circuit, stability, harmonics, etc.), user-friendly interface, strong industry support.

Ideal for: A wide range of industrial applications, from planning and design to operation and maintenance.

1.2.2 ETAP

Strengths: User-friendly interface, extensive library of components, suitable for both steady-state and dynamic analysis.

Ideal for: A broad range of industrial applications, particularly those requiring a balance between ease of use and comprehensive analysis capabilities.

DOI: 10.1201/9781003590514-1

1.2.3 MATLAB/Simulink®

Strengths: Highly flexible and customizable, excellent for research and development, powerful programming capabilities.

Ideal for: Advanced research, developing custom algorithms, and simulations that require high levels of customization.

1.2.4 PSCAD

Strengths: Specialized in electromagnetic transient (EMT) simulations, highly accurate for analyzing fast transients.

Ideal for: Analyzing the impact of switching operations, faults, and other transient events on industrial power systems.

1.2.5 EMTP

Strengths: Industry-standard software for EMT simulations, widely used and validated in the industry.

Ideal for: In-depth analysis of complex transient phenomena, particularly for critical infrastructure applications.

For a wide range of industrial power system analysis needs, **DIgital SImuLation of Electrical NeTworks (DIgSILENT) PowerFactory** and **ETAP** are often considered top choices due to their user-friendliness, comprehensive features, and industry acceptance. It's recommended to try out free trials or demos of at least these two software packages to determine which best suits your specific requirements.

1.3 Other Popular Programs

Here we briefly review some other power system softwares:

1.3.1 PowerWorld Simulator

PowerWorld Simulator is a powerful and user-friendly power system analysis software designed for simulating large high-voltage systems (up to 60,000 buses) over timeframes of minutes to days. Its interactive interface features animated one-line diagrams, color contours, and flexible display options. Beyond basic transmission line simulation, PowerWorld also handles Optimal Power Flow (OPF) and security-constrained OPF, making it suitable for power market simulations and educational purposes.

1.3.2 Simscape Electrical—MATLAB

Simscape Electrical (formerly SimPowerSystems and SimElectronics) simulates electrical power systems within the Simulink/MATLAB environment. It excels at modeling power generation, transmission, and distribution, particularly for control system design. Simscape Electrical helps you develop control systems and test system-level performance. You can parameterize your models using MATLAB variables and expressions, and design control systems for electrical systems in Simulink. You can integrate mechanical, hydraulic, thermal, and other physical systems into your model using components from the Simscape family of products. To deploy models to other simulation environments, including hardware-in-the-loop (HIL) systems, Simscape Electrical supports C-code generation.

1.3.3 WindMil

MilSoft's WindMil software is designed for efficient modeling and analysis of radial and loop electrical systems, specifically targeting utility engineers working on distribution system design and analysis at the feeder, substation, or system level. It features a graphical interface with editing tools, a single-line diagram with GIS integration through a LandBase tool, and a protective device database (LightTable) for protection coordination.

1.3.4 PSS®E

PSS/E is a powerful software suite developed by Siemens PTI that is used by utilities, grid operators, consultants, and researchers worldwide to simulate, analyze, and optimize the performance of power systems. It features an advanced one-line diagram with animation and loading graphs and has comprehensive analysis capabilities such as power flow, short circuit, stability (transient and dynamic), voltage stability, OPF, and more. PSS/E is also used for transmission planning and operation, distribution system analysis, renewable energy integration studies, market analysis, and grid modernization initiatives.

1.3.5 GE PSLF

GE's PSLF is a powerful suite capable of handling massive networks with up to 125,000 buses, organized using relational databases. The suite includes PSLF for load flow analysis with enhanced table and graphical navigation capabilities, SSTools for contingency analysis, PSDS for dynamic simulation offering both batch and interactive modes, and SCSC for robust short-circuit studies. It supports relational databases for data

preparation, editing, viewing, and reporting, and includes an Integrated
Development Environment (IDE) for Electric Power Command Language
(EPCL) scripting with advanced debugging. The dynamic simulation
features include plotting capabilities, and the short-circuit calculations
ensure comprehensive short-circuit study capabilities. Additionally, GE
offers stand-alone tools such as Distribution STAR (DSTAR) for distribution engineering, GE MAPS for hourly power market analysis and production cost comparisons, and GE MARS for multi-area system reliability
assessment. GE's PSLF software stands out with its updated user interface, seamless data integration, and robust analytical capabilities, ensuring thorough and reliable power system performance analysis.

1.3.6 PSAF

CYME International provides PSAF, a comprehensive power system analysis software package with integrated modules for various studies relevant to utilities and industrial applications. PSAF offers a suite of tools,
including CYMFLOW for power flow, motor starting, and AC contingency analysis; Short-circuit analysis (CYMFAULT) for short-circuit analysis; CYMHARMO for harmonic studies; CYMSTAB for transient stability
assessments; Line impedance & thermal analysis (CYMLINE) for one-line
diagram creation; WECS for wind energy system analysis; and CYME
programs for calculating line and cable parameters. Distribution system
analysis (CYMDIST) is a separate package designed for distribution network planning and analysis, offering several optional add-on modules.

1.3.7 ASPEN

Advanced Systems for Power Engineering (ASPEN) is an application with
a graphical interface designed for power system analysis. It includes integrated modules such as Power Flow for planning, design, and operational
studies of transmission, sub-transmission, and distribution networks;
OneLiner for short-circuit analysis and relay coordination, aimed at relay
engineers; a Relay Database serving as a central repository for relay information for utilities and industrial facilities; and DistriView, an integrated suite
for voltage drop, short-circuit calculations, and relay coordination specifically for distribution systems. ASPEN also offers data import/export capabilities and a built-in scripting language based on Beginner's All-purpose
Symbolic Instruction Code (BASIC). Additionally, ASPEN's OneLiner is
known for its speed, ease of use, and ability to handle networks with over
100,000 buses, while Power Flow offers an interactive graphics interface and
robust automatic-control algorithms, and DistriView supports voltage drop,
short circuit, relay coordination, harmonic analysis, and fault location.

1.3.8 EasyPower

EasyPower, offered by ESA, is a collection of software tools for power system analysis, design, measurement, and control, capable of simulating systems of any size. It provides a unified one-line diagram interface that incorporates diverse analysis functions. The EasyPower suite encompasses various modules, including EasyPower ShortCircuit for fault analysis, EasyPower PowerFlow for load flow studies, EasyPower PowerProtector for protective device coordination and database management, EasyPower Spectrum for harmonic and power quality analysis, EasyPower ArcFlash for arc flash hazard assessment, and EasyPower Measure for measurement analysis.

1.3.9 NEPLAN

NEPLAN Electricity is an intuitive and fully integrated software solution for analyzing power systems across transmission, distribution, and industrial networks. It supports functionalities including OPF, transient stability assessment, reliability analysis, harmonic analysis, dynamic simulation, and maintenance optimization. NEPLAN facilitates the analysis, planning, optimization, and management of power networks at all voltage levels, accommodating networks of any size. It offers a modular approach, based on international standards such as IEC, ANSI, and IEEE, to cater to both European and U.S. market needs. The software provides a comprehensive and user-friendly graphical interface with extensive libraries for network elements, protection devices, and control circuits, ensuring efficient and accurate study cases. The integration of GIS data and compatibility with cloud-based platforms further enhances its capabilities for modern power system analysis.

1.3.10 A Comparative Summary

In essence, PowerWorld and PSS/E are geared toward large-scale transmission systems, offering robust tools for detailed system planning and analysis. SimPowerSystems excels in control design and research, providing a flexible environment for developing and testing control strategies. WindMil is tailored for distribution systems, with robust GIS integration and protective device coordination tools. ASPEN and EasyPower provide practical, user-friendly analysis tools suitable for a range of applications, from short-circuit analysis to relay coordination and arc flash studies. Lastly, PSAF offers a balanced suite with specialized modules for power flow, short circuit, harmonic analysis, and transient stability assessments, making it a versatile tool for both utilities and industrial applications.

1.4 Why DIgSILENT PowerFactory for This Book?

DIgSILENT PowerFactory is widely regarded as one of the leading power system analysis software platforms due to its comprehensive capabilities, flexibility, and user-friendly interface. It supports a wide range of simulation types, making it suitable for almost all aspects of power system analysis. PowerFactory can handle complex, large-scale system models across various domains of power engineering. The application can practically meet all the requirements of a real network, including quasi-dynamic simulation, cable reinforcement, open-tie optimization, reliability, and probabilistic analysis, contingency analysis, transient stability analysis, transfer capacity analysis, protection coordination, unit commitment, RMS/EMT simulation, network reduction, arc flash analysis, harmonic analysis, hydropower plant analysis, wind farm and power park energy analysis, reactive power capability assessment, HVDC and railway systems analysis, switching and lightning transients analysis, state estimation, optimal power restoration, generation adequacy analysis, and modal/eigenvalue analysis. As you can see, the capabilities of DIgSILENT PowerFactory are extensive, making it impossible to cover everything in a single book. The remainder of this book will focus on and examine only the most significant simulations and analyses. Each chapter will explain fundamental concepts and provide instructions on how to simulate these key aspects effectively.

The DIgSILENT PowerFactory software is used for power network analysis, as confirmed by the authors' comparison, qualifications, and experiences. For more detailed work and further work, you can refer to the references [1–6] at the end of this chapter.

1.5 Summary

Selecting the best power system analysis software is a complex process that depends on various factors. Based on the authors' experience, DIgSILENT PowerFactory is recommended for power network analysis in this book. This software offers numerous advantages, making it a specialized and highly effective tool for power system analysis, as corroborated by independent research and expert reviews.

References

1. Bam, L., Jewell, W. "Review: Power system analysis software tools," *IEEE Power Engineering Society General Meeting, 2005*, San Francisco, CA, USA, 2005, vol. 1, pp. 139–144, https://doi.org/10.1109/PES.2005.1489097.
2. Eidiani, M., Zeynal, H., Zakaria, Z. "A comprehensive study on the renewable energy integration using DIgSILENT," *2023 IEEE 3rd International Conference in Power Engineering Applications (ICPEA)*, Putrajaya, Malaysia, 2023, pp. 197–201, https://doi.org/10.1109/ICPEA56918.2023.10093153.
3. Eidiani, M. "Modeling renewable energy resources using DIgSILENT PowerFactory software," In: Chenniappan, S., Padmanaban, S., Palanisamy, S. (eds) *Power Systems Operation with 100% Renewable Energy Sources*, 2024, pp. 165–202. https://doi.org/10.1016/B978-0-443-15578-9.00013-3.
4. Parhamfar, M., Eidiani, M., Abtahi, M. "Distributed energy storage system: Case study," In: Palanisamy, S., Chenniappan, S., Padmanaban, S. (eds) *Distributed Energy Storage Systems for Digital Power Systems*, Elsevier, 2025, pp. 395–422. https://doi.org/10.1016/B978-0-443-22013-5.00013-7
5. Karlsson, B. "Comparison of PSSE & PowerFactory." Thesis, Uppsala universitet, Elektricitetslära, 2013, https://patents.google.com/scholar/15337555155925632399.
6. Kumar, Y., Devabhaktuni, V.K., Vemuru, S. "Comparison of power system simulation tools with load flow study cases," *2015 IEEE International Conference on Electro/Information Technology (EIT)*, Dekalb, IL, USA, 2015, pp. 290–294, https://doi.org/10.1109/EIT.2015.7293355.

2

Power System Modeling

2.1 Introduction

Engineers must collect specific data and take into account a number of important factors in order to create an accurate model of a power grid. They first require a comprehensive blueprint of the physical layout of the grid, which includes details about the distribution points, transformers, transmission lines, and power plants. An accurate representation of all involved grid components is required, including information on generator characteristics, transformer voltage, and power ratings, and transmission line series resistance and reactance, even in some studies shunt conductances.

It is essential to comprehend the kinds, quantities, and locations of electricity demand. This covers loads from homes, businesses, and industries. It's also crucial to know the types, sizes, and locations of power plants, as well as those that use conventional and renewable energy sources such as wind and solar. It is crucial to model the control systems that manage the flow of power and voltage in the grid. This can be devices such as power system stabilizers (PSS) and automatic voltage regulators (AVRs). Potential disruptions like short circuits or abrupt changes in load must also be taken into account, as well as the grid's possible reaction.

Engineers can develop trustworthy models that aid in performance analysis, grid optimization, and future growth planning by carefully taking these variables into account and gathering precise data.

The appendix provides instructions on how to utilize the software and the potential for more work. You can consult the references [1–12] at the conclusion of this chapter for further in-depth work and additional research.

2.2 Key Points in Working with Power Systems Programs

The majority of power system analysis programs, including DIgSILENT, benefit from the recommendations made in this section. Because of this, mastering one software can significantly simplify the process of learning another one.

DOI: 10.1201/9781003590514-2

1. **Windows Structure:** Software consists of different windows and in different layers. You can easily move windows or hide layers.

2. **Hold the Mouse:** By holding the mouse over each element, you will see additional explanations, such as finding the Load Flow icon.

3. **One-Click:** Clicking on any element selects it so that activities can be carried out on it at a later time without requiring any particular action at the moment.

4. **Click Again:** By clicking again and holding the mouse on the selected element, you can move the element.

5. In actuality, you must choose any part in order to move it (point 3). The element can then be moved by holding down the mouse button while you click again after releasing it (point 4).

6. **Double-Click:** By double-clicking, you can enter the element data set and change its information.

7. By double-clicking on the keys (Cubicle-Switch) of each element, it can be opened and closed (Figure 2.1).

8. Result Boxes are automatically created and can be moved or hidden (Figure 2.1).

9. Uppercase and lowercase letters are different in names such as MATLAB software. For example, (BUS) and (Bus) and (bus) are three different names. Also, names are case sensitive, for example, (Bus4) and (Bus 4) are different.

10. Every graphic program, including Photoshop, allows you to apply various layers to every image. Each of these layers enhances the original image with a different type of graphic or information. You can see multiple layers simultaneously if the grid is tiny. Nevertheless, displaying all the superfluous layers will clog the image, if the grid is large. An illustration of various graphic layers is presented in Figure 2.2.

FIGURE 2.1
Result boxes and cubicle.

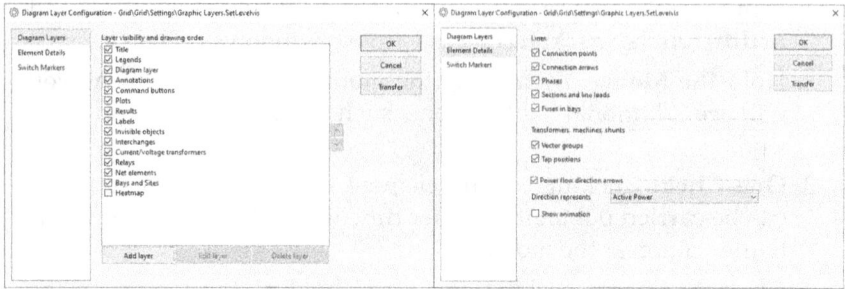

FIGURE 2.2
Layers window.

11. If you wish to remove the most recent simulation results from the network, you must press the reset button after every load flow, short circuit, stability analysis, etc.

2.3 Saving Graphic Output Pages

Most professional power system software uses a variety of common formats for their graphic output, such as .bmp, .emf, .gif, .jpg, .pdf, .png, .svg, .tif, .wmf. Here is a summary of the key features of each format:

A. Raster Image Formats

A1. **BMP:** A simple, uncompressed format, often resulting in large file sizes. It's suitable for high-quality images but less efficient for web use.

A2. **GIF:** Supports animation and transparency, but is limited to 256 colors. Ideal for simple graphics and animations.

A3. **JPEG/JPG:** A popular format for photographs due to its high compression ratio, often at the cost of some image quality.

A4. **PNG:** Offers lossless compression, making it suitable for images with sharp edges and text. It also supports transparency and multiple color depths.

A5. **TIFF:** A versatile format that supports lossless compression, high color depth, and multiple layers. Commonly used for professional printing and image editing.

B. Vector Image Formats

    B1. **SVG:** A scalable vector graphic format, meaning images can be resized without losing quality. It's ideal for logos, icons, and illustrations.

    B2. **WMF:** A Windows Metafile format, used for vector graphics and simple images. It's often used for Windows-specific graphics.

    B3. **EMF:** Enhanced Metafile format, a more advanced version of WMF, offering better color depth and support for more complex graphics.

What are the main distinctions between raster (A) and vector (B)? Vector graphics are scalable without sacrificing quality since they are determined by mathematical formulae. Raster graphics are made up of pixels, and enlarging them can result in pixelation. For this reason, unless it is required, we attempt to make all outputs vector. But vector files are much harder to edit than raster files. See Figure 2.3. We compared a few examples of zoomed photos.

Lastly, we want to highlight three points:

Note 1: Note that .wmf files are not supported by MS-Paint. If opened with this program, they are first converted to raster format and then displayed.

Note 2: You can view .wmf files with ACDSee or add them to MS-Word and view them.

Note 3: If you use the Print Screen (Prt Sc) key for output shapes, the images are still saved in memory as raster.

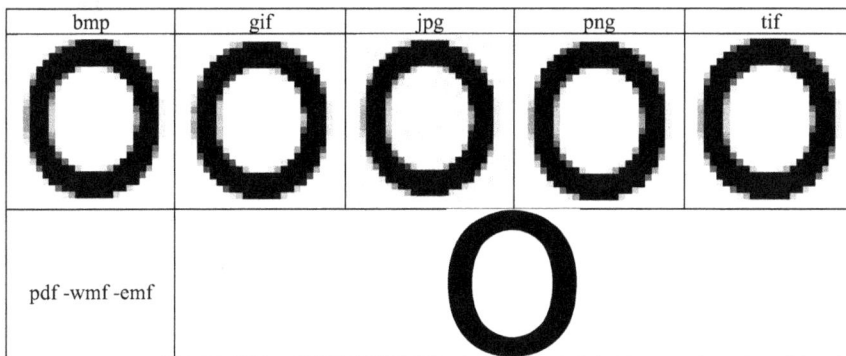

bmp	gif	jpg	png	tif
O	O	O	O	O
pdf -wmf -emf		O		

FIGURE 2.3
A simple comparison between different image formats (Vector and Raster.)

2.4 Building a Simple 9-Bus Network

The drawing toolbox, which opens next to the main screen (left or right), is required in order to create any network. See Figure 2.4. All the necessary components for power system modeling are included in this toolbox. More simulation-related features are included in each software update. Simply move the mouse pointer over each element to see its name so you can become familiar with it.

To build a network, you should pay attention to the following points:

1. The busbar must be allocated first when constructing a network. After that, other components like transformers, lines, and loads can be added.

2. Once an element has been chosen and positioned on the page, you can click and select it to move, rotate, resize, or remove it. Your keyboard's Delete key is also functional. In software, the delete sign is typically represented by a cross or trash can key.

3. The software selects a name for each element after it has been placed on the page. Simply click on the element preselected name to make the name modification.

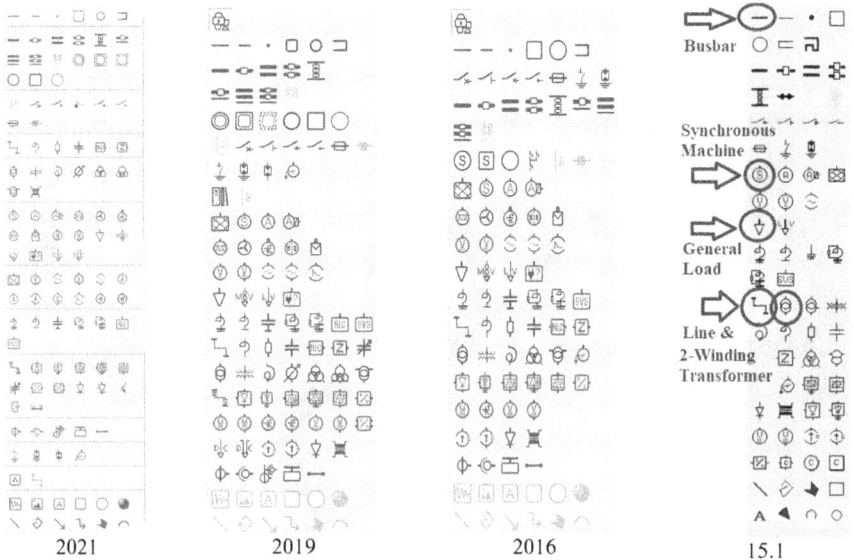

2021   2019   2016   15.1

FIGURE 2.4
Elements toolbox for drawing (in different editions.)

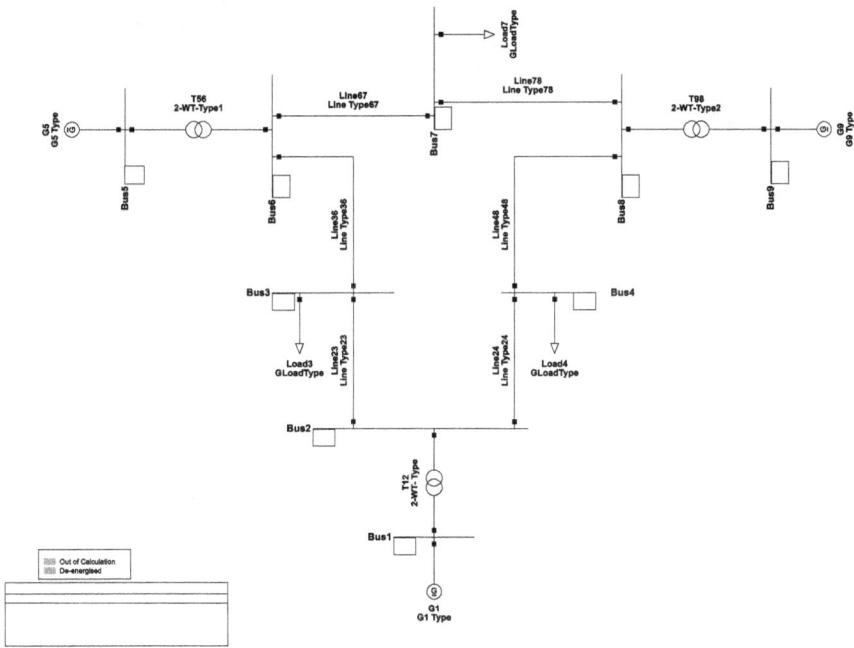

FIGURE 2.5
Basic drawing of a 9-bus network.

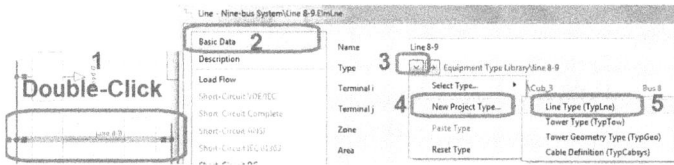

FIGURE 2.6
Determining the transmission line type.

4. Next, attempt to build your network precisely like the one shown in Figure 2.5. Take note of the word precisely!

5. A new type must be defined for all elements. To do this, proceed as shown in Figure 2.6.

6. Double-click (No. 1) on the element in the Basic Data page (No. 2), click the black downward arrow (No. 3) in front of Type, and select the New Project Type option (No. 4). Then, in front of Name, enter the desired type name (for example, element name+type) and click on OK. Do this only once for each element.

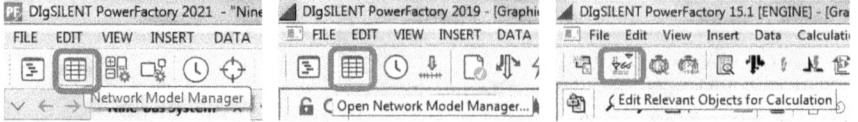

FIGURE 2.7
Key to enter network element information in different versions.

7. If a type is defined more than once for each element, the software will enter (1) or (2) and ... for subsequent names, and you must enter the Select Project Type section to delete them and delete the additional types.
8. To enter element information values, you must use the (Edit Relevant Objects for Calculation) or (Network Model Manager) option located at the top of the main menu (Figure 2.7).
9. The network's primary components are indicated in green in Figure 2.8, while each one's type is indicated in red. You can enter the network-related data by clicking on each of them.

The necessary data for the elements is then entered in accordance with the recommended guidelines and tables.

2.4.1 Entering Element Information

According to the data at the top of each table, the necessary information is shown in red for types and green for each group of elements in this section. For instance, as illustrated in Figure 2.9, you must first navigate to the busbars section (green, No. 2) before selecting the (Basic data) sub-window (No. 3) to access the information (Basic Data for Busbar). See Table 2.1.

2.4.1.1 Information about Loads

You may have three types of loads instead of one, like busbars. Enter the information similarly. See Table 2.2.

2.4.1.2 Information about Lines

Enter the resistance, reactance, and voltage information of the lines. Assume the length of the lines is 1 km. See Table 2.3.

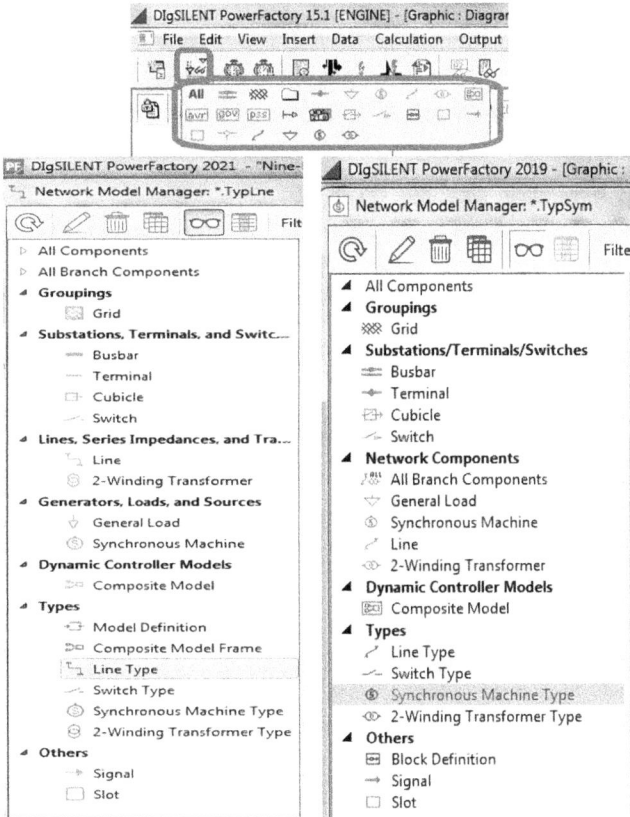

FIGURE 2.8
Complete information about elements and their types in different versions (2019 and 2021).

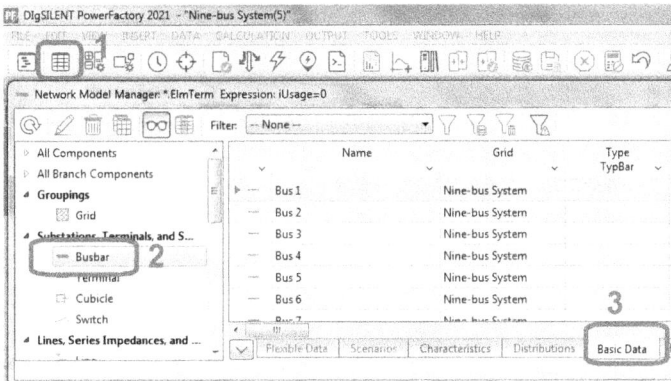

FIGURE 2.9
Directions to (Basic Data for Busbar.)

TABLE 2.1

Busbar Information Entry

Basic Data for Busbar					Basic Data for Busbar Type		
> All Branch Components	Name	Grid	Type TypBar	Nom.L-L Volt. kV			kV
∨ Groupings							
Grid	Bus1	Grid	Busbar Type1	16.5			
∨ Substations, Terminals...	Bus2	Grid	Busbar Type	230.	∨ Types	Busbar Type1	16.5
Busbar	Bus3	Grid	Busbar Type	230.	Busbar Type	Busbar Type	230.
Terminal	Bus4	Grid	Busbar Type	230.	Line Type	Busbar Type5	18.
Cubicle	Bus5	Grid	Busbar Type5	18.	General Load Ty...	Busbar Type9	13.5
Switch	Bus6	Grid	Busbar Type	230.	Synchronous M...		
∨ Lines, Series Impedan...	Bus7	Grid	Busbar Type	230.	Results		
Line	Bus8	Grid	Busbar Type	230.		Basic Data	Description	Version
2-Winding Tran...	Bus9	Grid	Busbar Type9	13.5			
General Load							
Synchronous M...	Basic Data	Description	Load Flow	Short-Circ			

TABLE 2.2

Load Information Entry

Load Flow Data for Loads					Load Flow for Loads Type			
2-Winding Tran...	Name	Grid	Act.Pow. MW	React.Pow. Mvar	∨ Types		e_cP	e_cQ
Generators, Loads, an...					Busbar Type	Filter: -- None --		
General Load	Load3	Grid	125.	50.	Line Type	Name		
Synchronous M...	Load4	Grid	90.	30.	General Load Type	GLoadType	1.6	1.8
Load Flow Controller ...	Load7	Grid	100.	35.	Synchronous Machine Type			
Line type					2-Winding Transformer Type			
General Load Ty...					Results			
	Description	Load Flow	Short-Circuit VDI			Version	Load Flow	Short-Circuit VDE/IEC

TABLE 2.3

Line Information Entry

Basic Data for Line Types							Load flow for Line Types			
∨ Load Flow Controll...	Name	Rtd.Volt. kV	Rtd.Current kA	R (AC,20°C) Ohm/km	X' Ohm/km		∨ Load Flow Controll...	Name	B' uS/km	
Station Con...							Station Con...			
∨ Types	Line Type23	230.	1.	5.29	44.965		∨ Types	Line Type23	120.	
Busbar Type	Line Type24		1.	8.993	48.668		Busbar Type	Line Type24	120.	
Line Type	Line Type36	230.	1.	16.928	85.169		Line Type	Line Type36	150.	
General Loa...	Line Type48	230.	1.	20.631	89.93		General Loa...	Line Type48	150.	
Synchronous...	Line Type67	230.	1.	4.4965	38.088		Synchronou...	Line Type67	169.	
2-Winding T...	Line Type78	230.	1.	6.2951	53.3232		2-Winding T...	Line Type78	169.	
∨ Others							∨ Others			
Results	Basic Data	Description	Version	Load Flow	Short-Circuit VDE/IEC	Sho	Results	Load Flow	Short-Circuit VDE/I	

2.4.1.3 Information about Transformer

In order to enter transformer information, double-click the transformer and select the (Flip Connections) button if the HV-LV transformer buses have been moved. The HV-LV bus row ought to be accurate. See Table 2.4.

2.4.1.4 Information about Generators

First, enter the information in Table 2.5 and then enter the information in RMS Table 2.6.

TABLE 2.4

Transformer Information Entry

Basic Data for Transformer

	Name	Type TypTr2	HV-Side	LV-Side
T12	2-WT- Type	Bus2	Bus1
T56	2-WT-Type1	Bus6	Bus5
T98	2-WT-Type2	Bus8	Bus9

Basic Data   Description   Load Flow   Short-Circuit VDE/IEC

Basic Data for Transformer Types

	Name	rtd.Pow. MVA	HV-rtd.Volt. kV	LV-Rtd.Volt. kV	Shc Volt. %	HV-Vec.Grp.	LV-Vec.Grp.	Ph.Shift *30deg	Name	uk0 %
2-WT- Type	250.	230.	16.5	14.4	YN	D	5.	YNd5	3.
2-WT-Type1	200.	230.	18.	12.5	YN	D	5.	YNd5	3.
2-WT-Type2	150.	230.	13.8	8.79	YN	D	5.	YNd5	3.

Basic Data Description Version Load Flow Short-Circuit VDE/IEC Short-Circuit Complete Short-Circuit ANSI Short-Circuit IEC 61363 Short-

TABLE 2.5

Generators Information Entry

Basic Data for Synchronous Machine Types

	Name	App.Pow. MVA	Rtd.Volt. kV	Pow.Fact.	No. of Phases	Connection
G1 Type	247.5	16.5	1.	3	YN
G5 Type	192.	18.	0.85	3	YN
G9 Type	128.	13.8	0.85	3	YN

Load flow Data for Synchronous Machine

Name	Spinnin	Ref.Machine	Local Controller	Act.Pow. MW	React.Pow. Mvar	Pow.Fact.	Voltage p.u.	Angle deg	Max.Act.. MW	Pr(rated) MW	Pmax MW	Rating F..	Pr(rated) MW
G1	☐	☑	constv	0.	0.	1	1.04	0.	9999.	247.5	247.5	1.	247.5
G5	☐	☐	constv	140.	50.	0.9417419	1.025	0.	9999.	163.2	163.2	0.99999	163.2
G9	☐	☐	constv	85.	20.	0.9734172	1.025	0.	9999.	108.8	108.8	1	108.8

TABLE 2.6

Generators RMS Information Entry

Name	Tag[Pgn] s	rstr p.u.	xl p.u.	Type	Td' s	Tq' s	Td'' s	Tq'' s	xd p.u.	xq p.u.	xrld p.u.	xrlq p.u.	xd' p.u.
G1 Ty...	9.55	0.	0.083	0	3.73	0.	0.05	0.05	0.36	0.24	0.	0.	0.15
G5 Ty...	4.165	0.	0.141	1	0.8	0.12	0.05	0.05	1.72	1.66	0.	0.	0.23
G9 Ty...	2.765	0.	0.0949	1	0.806	0.12	0.05	0.05	1.68	1.61	0.	0.	0.23

xq' p.u.	xd'' p.u.	xq'' p.u.	x0 p.u.	r0 p.u.	Main Flux Sat.	Saturation	SG10 p.u.	SG12 p.u.	x2 p.u.	r2 p.u.
0.3	0.1	0.1	0.1	0.	Quadratic (S...	d- and q...	0.13	0.32	0.2	0.
0.378	0.2	0.2	0.1	0.	Quadratic (S...	d- and q...	0.13	0.32	0.2	0.
0.32	0.2	0.2	0.1	0.	Quadratic (S...	d- and q...	0.13	0.32	0.2	0.

Next, select the RMS Synchronous Machine Type tab (red) and enter the following information.

2.5 Load Flow

To obtain the load flow from the network after filling out the previous section, click the **Calculate Load Flow** icon. Figure 2.10 should be your result. You can use the file Chapter2.pfd to continue working.

2.6 Two-Choice Questions (Yes/No)

1. Power system network modeling is essential for analyzing the behavior of power systems.
2. Impedance values of transmission lines and transformers are important parameters in power system models.
3. Load flow analysis does not require accurate network modeling.
4. Network topology is not a critical factor in power system modeling.
5. The power system network model is static and does not change over time.
6. Load flow analysis is used to determine the steady-state operating conditions of a power system.
7. Short-circuit analysis requires detailed network models.
8. Transient stability analysis requires detailed dynamic models of generators and loads.
9. The power system network model does not need to consider the impact of renewable energy sources.
10. Short-circuit analysis is used to determine the steady-state operating conditions of a power system.
11. The network model does not need to consider the impact of control systems.
12. The network model can be used to analyze the impact of different operating scenarios.
13. The network model is only used for offline analysis.
14. The network model can be used for real-time applications, such as state estimation.
15. The accuracy of power system simulations depends on the quality of the network model.
16. The network model is always a complete representation of the physical power system.

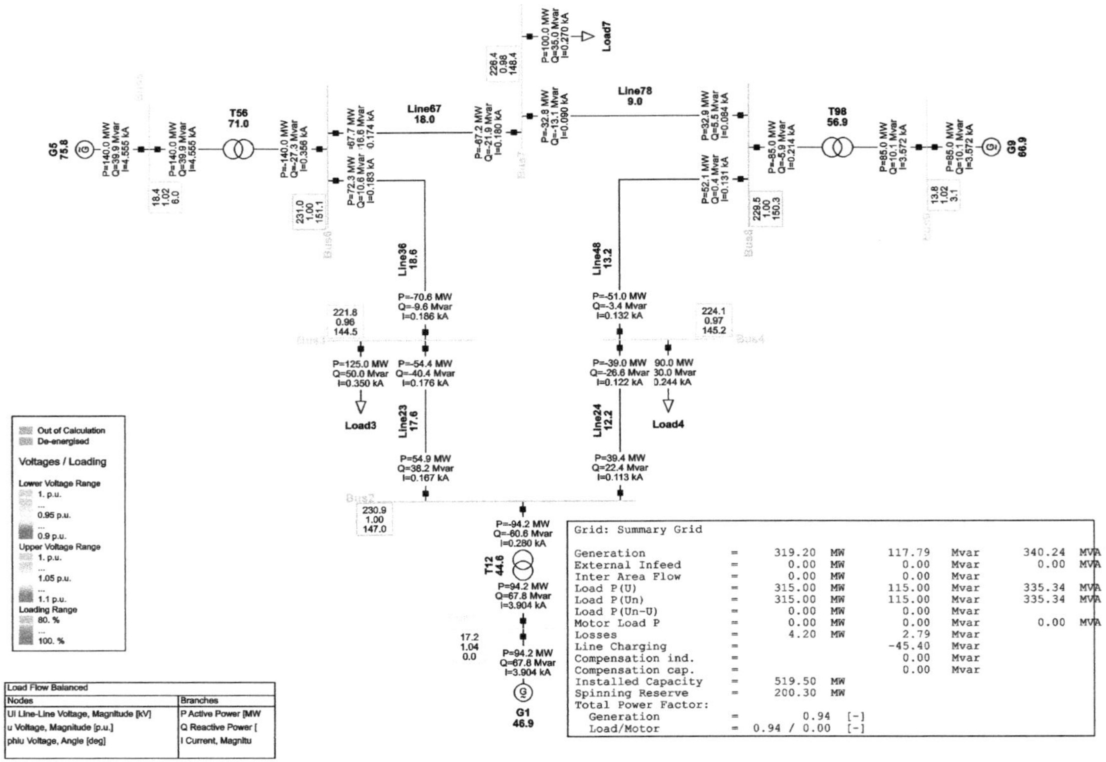

FIGURE 2.10
Power flow result of the 9-bus network.

17. Simplifications and approximations are often used in network modeling.
18. The network model can be used to assess the impact of different protection schemes.
19. The network model is not used for the planning and design of power systems.
20. The network model does not need to consider the impact of disturbances, such as faults.

2.6.1 Key Answers to Two-Choice Questions

Yes	1, 2, 6–8, 12, 14, 15, 17, 18
No	3–5, 9–11, 13, 16, 19, 20

2.7 Summary

Modeling a power network is the cornerstone of effective network analysis, as the accuracy of these models significantly impacts simulation results. Ongoing research aims to refine models of power system elements to enhance their precision, recognizing that inaccurate models can lead to flawed analyses. The integration of new elements, such as renewable energy systems, further underscores the need for accurate and cohesive modeling.

References

1. Eidiani, M., Zeynal, H., "New approach using structure-based modeling for the simulation of real power/frequency dynamics in deregulated power systems," *Turkish Journal of Electrical Engineering and Computer Sciences,* 2014, 22(5), pp. 1130–1146, https://doi.org/10.3906/elk-1208-90.
2. Eidiani, M. "Modeling renewable energy resources using DIgSILENT PowerFactory software," In: Chenniappan, S., Padmanaban, S., Palanisamy, S. (eds) *Power Systems Operation with 100% Renewable Energy Sources,* Elsevier, 2024, pp. 165–202. https://doi.org/10.1016/B978-0-443-15578-9.00013-3
3. Eidiani, M., Rouzbehi, K., *Fundamentals of Power System Transformers Modeling, Analysis, and Operation,* Taylor & Francis Group, CRC Press, 2025, pp. 1–127. http://www.routledge.com/9781032881751

4. Eidiani, M., Rouzbehi, K., *Advanced Topics in Power Systems Analysis: Problems, Methods, and Solutions,* Taylor & Francis Group, CRC Press, 2024, pp. 1–120. http://www.routledge.com/9781032828664

5. Eidiani, M., Heidari, V., *Fundamentals of Power Systems Analysis 1: Problems and Solutions,* Taylor & Francis Group, CRC Press, 2023, pp. 1–215, https://doi.org/10.1201/9781003394433.

6. Eidiani, M., "An efficient differential equation load flow method to assess dynamic available transfer capability with wind farms," *IET Renewable Power Generation,* 2021, 15, pp. 3843–3855, https://doi.org/10.1049/rpg2.12299.

7. Eidiani, M., "A new load flow method to assess the static available transfer capability," *Journal of Electrical Engineering and Technology,* 2022, 17(5), pp. 2693–2701, https://doi.org/10.1007/s42835-022-01105-3.

8. Rouzbehi, K., Zhang, W., Candela, J., Luna, A., Rodriguez, P., "Generalized voltage droop strategy for power synchronization control in multi-terminal DC grids-an analytical approach," *2015 International Conference on Renewable Energy Research and Applications,* 2015, pp. 1568–1574, "http://dx.doi.org/10.1109/ICRERA.2015.7418670"10.1109/ICRERA.2015.7418670.

9. Karami, E., Madrigal, M., Gharehpetian, G. B., Rouzbehi, K., Rodriguez, P., "Single-phase modeling approach in dynamic harmonic domain," *IEEE Transactions on Power Systems,* vol. 33, no. 1, pp. 257–267, 2018.

10. Rouzbehi, K., Miranian, A., Candela, J. I., Luna, A., Rodriguez, P., "Intelligent voltage control in a DC micro-grid containing PV generation and energy storage," *2014 IEEE PES T&D Conference and Exposition,* pp. 1–5, 2014.

11. Martino, M., Citro, C., Rouzbehi, K., Rodriguez, P., "*Efficiency analysis of single-phase photovoltaic transformer-less inverters,*" RE&PQJ 10 (12), 2012.

12. Karami, E., Gharehpetian, G. B., Rouzbehi, K., CheshmehBeigi, H. M., "A generalized representation of VSC-HVDC based AC/DC microgrids for power flow studies," *2018 Smart Grid Conference* (SGC), pp. 1–5, 2018.

3

Load Flow Control Methods

3.1 Introduction

This chapter demonstrates load flow[1] (power flow) control methods using DIgSILENT PowerFactory software, providing essential information for power system analysis.

Load flow analysis is a fundamental tool in power system engineering that determines the steady-state operating conditions of a network. This analysis calculates critical parameters including voltage magnitudes, phase angles, and both active and reactive power flows through transmission lines and generators. It serves as a cornerstone for power system planning, operation, and control, enabling engineers to evaluate and optimize network performance.

Power system planning uses load flow analysis to find the best size and dimensions for new generating and transmission infrastructure, evaluate the effects of load variations, and spot any bottlenecks. Load flow analysis aids system operation by tracking the power system's current condition in real time, spotting possible issues, and optimizing power flow to reduce losses and boost effectiveness.

To maintain system stability and efficiency, a variety of control techniques are employed. Devices used in voltage control include voltage regulators, tap-changing transformers, and static VAR compensators. Reactive power control uses techniques including reactor banks, capacitor banks, generator excitation control, and FACTS devices to maintain reactive power balance. Automatic generation control (AGC) maintains system frequency while economic dispatch optimizes power generation allocation in real power control.

The power flow equations are solved using a variety of techniques, such as the **Newton–Raphson** approach, the **Gauss–Seidel** method, and the **fast-decoupled load flow method**. Every approach has pros and cons, and the particular application and system properties determine which approach is best.

Power system operators maintain grid reliability and efficiency through the strategic application of load flow analysis and control techniques.

The appendix provides instructions on how to utilize the software and the potential for more work. You can consult the references [1–9] at the conclusion of this chapter for further in-depth work and additional research.

DOI: 10.1201/9781003590514-3

3.2 Normal AC Load Flow

The 9-bus network shown in Figure 3.1 serves as our example for AC load flow analysis. Table 3.1 presents the Total System Summary data, which can be viewed in both graphical and textual formats. The network configuration and load flow results will be detailed in the following sections.

3.2.1 Coloring

The network's performance can be quickly assessed through color-coded visualizations. As shown in Figures 3.1 and 3.2, green indicates nominal voltage (1 per unit), while cool blues represent lower voltages and warm reds indicate higher voltages or elements loaded above 80%.

3.2.2 Vector Diagram

A "**Vector Plot**" or "**Vector Diagram**" shows the relationship between each line's voltage and current. Figure 3.3 shows the voltage and current at the two ends of the line along with their magnitude and angle. The current in each line is always thought to be toward the inside of the line, which explains why the current angle at the line's beginning and end is 180° different.

3.2.3 PV and QV Curves

Voltage stability is a critical aspect of power system analysis. The basic assessment method involves monitoring bus voltage while incrementally increasing load demand (active and reactive power). The voltage stability limit—or voltage collapse point—is reached when the load flow fails to converge. This analysis can be performed by varying reactive power, active power, or both simultaneously.

The relationship between voltage and power is illustrated through two characteristic curves: the PV curve (Figure 3.4) and QV curve (Figure 3.5), which demonstrate the maximum allowable active and reactive power loads for stable operation. Figure 3.6 demonstrates a load flow scenario approaching the stability limit, where the color-coded visualization clearly shows elevated element loading and reduced bus voltages.

3.2.4 Loading Lines and Bar Diagram

Line current curve plots provide rapid identification of overloaded transmission lines. The line current curve plot also identifies transmission lines with maximum and minimum loading levels. Figure 3.7 illustrates this through a plot of the nine-bus network.

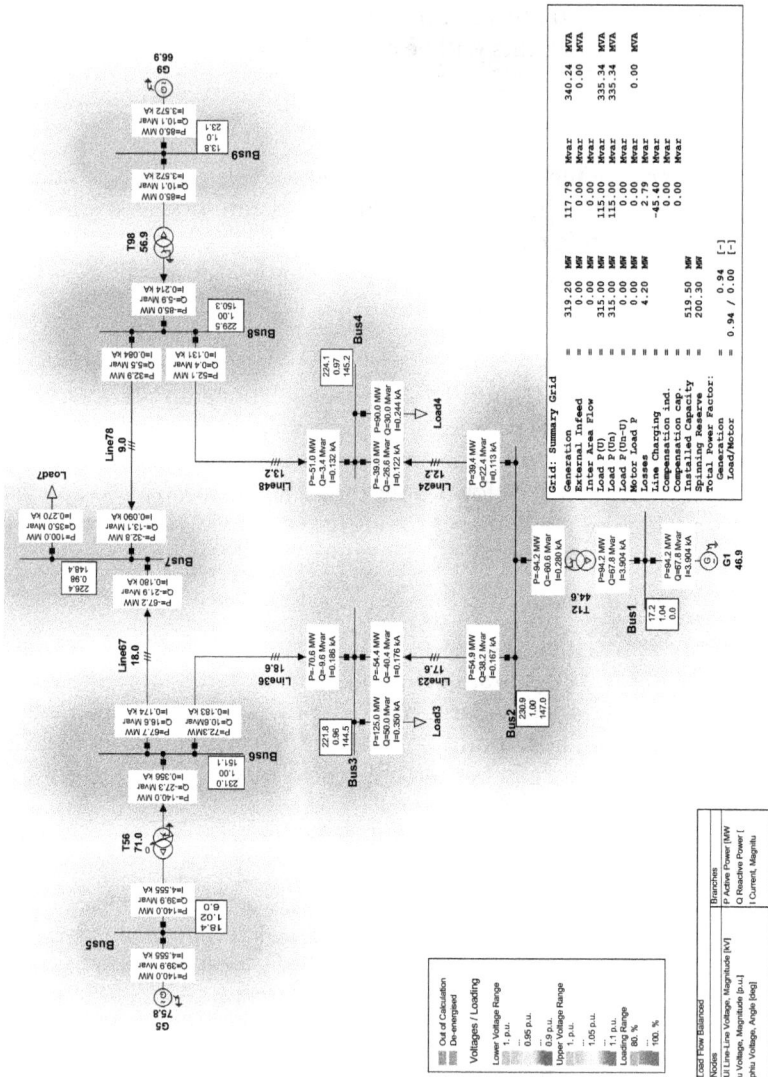

FIGURE 3.1
The 9-bus network load flow.

TABLE 3.1

Total Systems Summary

Generation = 319.20 MW 117.79 Mvar 340.24 MVA	Generation
External Infeed = 0.00 MW 0.00 Mvar 0.00 MVA	Power received from external network
Load P(U) = 315.00 MW 115.00 Mvar 335.34 MVA	Load at load distribution voltage
Load P(Un) = 315.00 MW 115.00 Mvar 335.34 MVA	Load at rated voltage
Load P(Un-U) = 0.00 MW 0.00 Mvar	Load difference in the two above cases
Motor Load = 0.00 MW 0.00 Mvar 0.00 MVA	Motor load
Grid Losses = 4.20 MW 2.79 Mvar	Active and reactive losses
Line Charging = -45.40 Mvar	Reactive power generated by the line
Compensation ind.= 0.00 Mvar	Reactor installed in the network
Compensation cap.= 0.00 Mvar	Capacitor installed in the network
Installed Capacity = 519.50 MW	Total installed capacity of generators
Spinning Reserve = 200.30 MW	Spin reserve i.e. the amount of reserve for overproduction

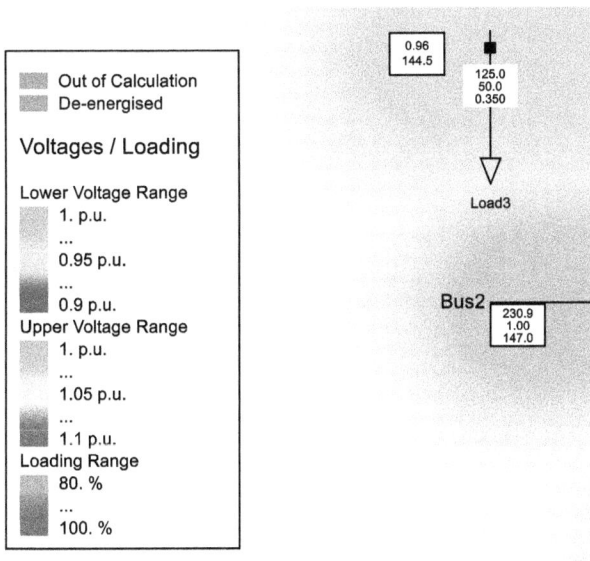

FIGURE 3.2
Coloring based on voltage and loading.

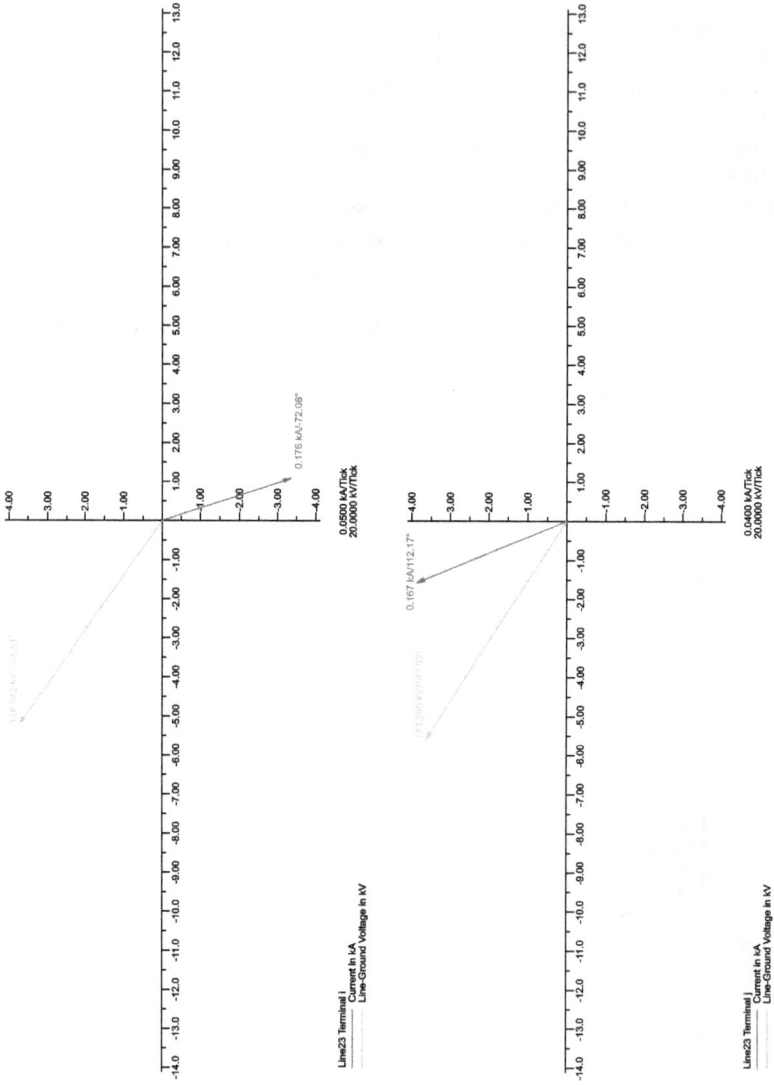

FIGURE 3.3
Output of the vector diagram of lines 2–3.

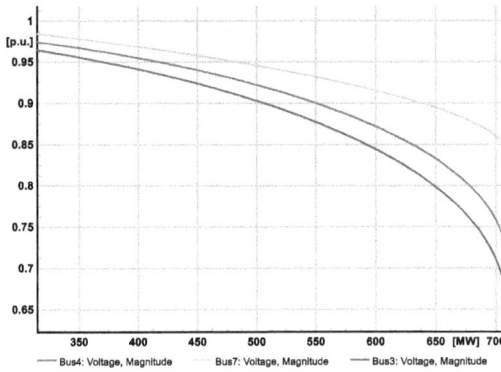

FIGURE 3.4
PV curve for load buses.

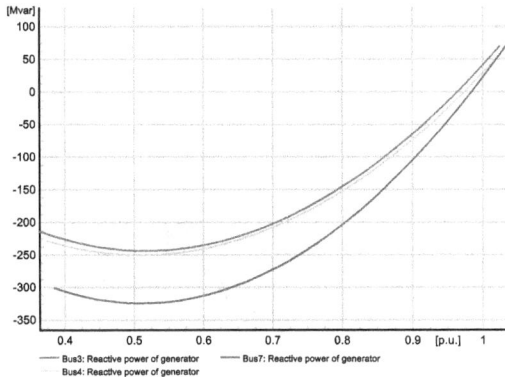

FIGURE 3.5
QV curve for load buses.

The voltage bar diagram also plots the voltage of all buses per unit (Figure 3.8). This curve visually represents the voltage levels of buses, indicating whether they are below or above the nominal voltage (typically 1 per unit).

3.2.5 Reactive Power Compensation with Capacitor

Adding a capacitor to the network is the most straightforward method of improving voltage, reducing line overload, and compensating for reactive power. Since the majority of loads are inductive, installing a capacitor helps with some of the issues brought on by reactive power. Capacitors have two drawbacks: they are discrete components, meaning they need to be physically installed and connected to a circuit, and can only provide reactive power. On the plus side, they are inexpensive and simple to install.

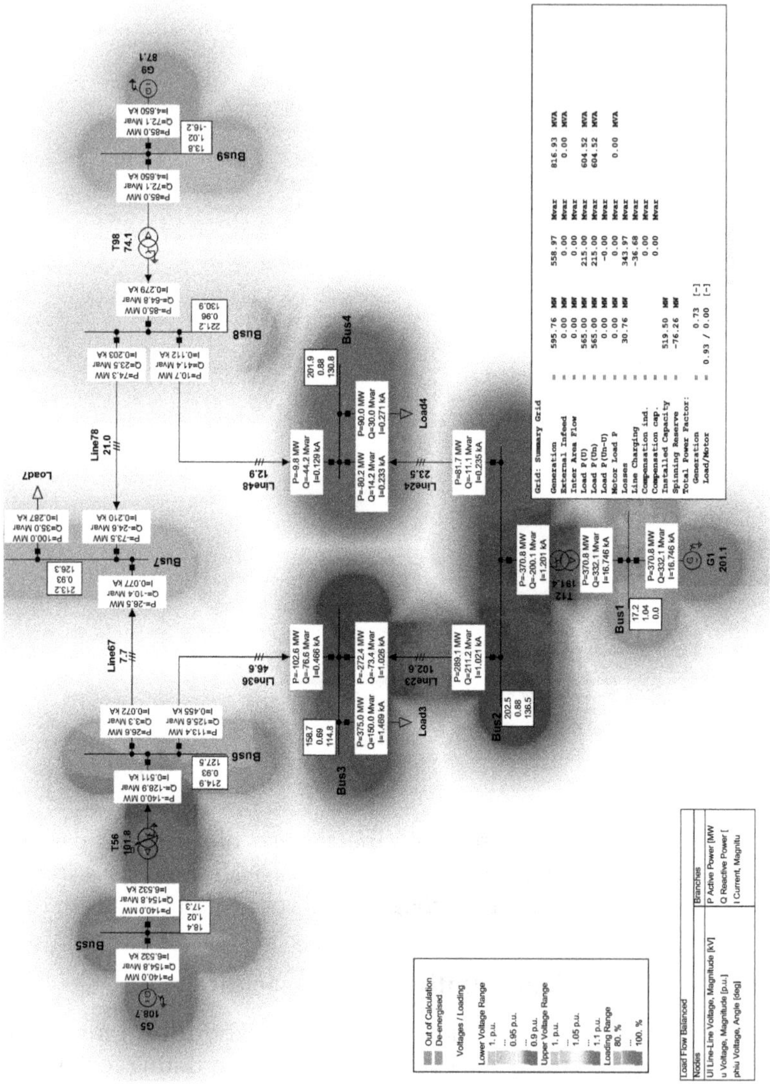

FIGURE 3.6
The load flow near the stability limit.

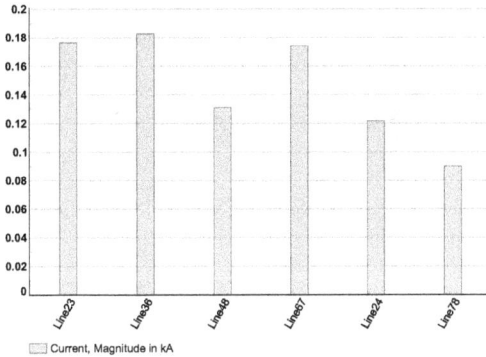

FIGURE 3.7
The line current curve plot.

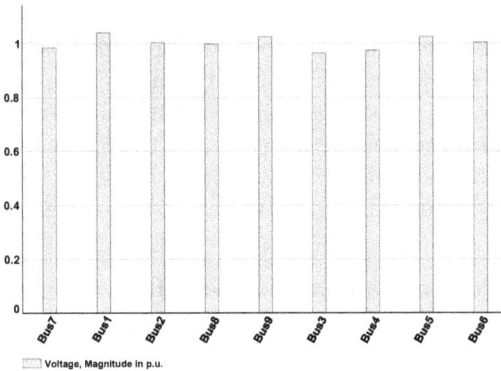

FIGURE 3.8
An example of a bar diagram.

Refer to Figure 3.9. Without a capacitor, the busbar voltage 3 in case (a) is 0.96 pu. When a 40 MVAr capacitor is used in case (b), the busbar voltage rises to 1 pu. The losses in cases (a) and (b) are 4.2 MW+j 2.79 MVAr and 3.95 MW+j –2.21 MVAr, respectively. It can be seen that as the voltage increases, the losses decrease.

3.2.6 Consider Voltage Dependency of Loads

We are aware that the loads used determine the voltage at any given location in the network. A capacitor, as a source of reactive power, can be used to raise the voltage because loads are often inductive and use reactive power, which results in a voltage drop.

We remind you in this section that voltage also affects loads. For instance, a standard 100W, 220V bulb operates at its rated power (100W) when supplied

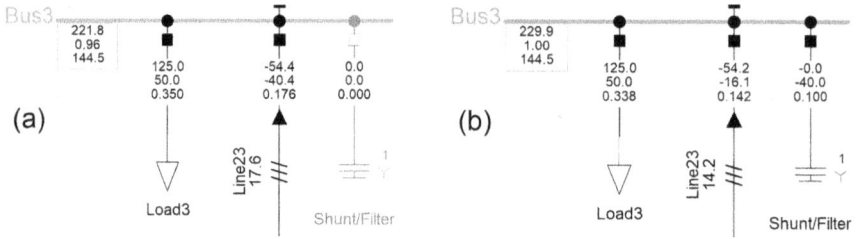

FIGURE 3.9
Capacitor effect on bus voltage.

with its nominal voltage (220V). A decrease in lamp voltage results in a decrease in lamp power consumption, while an increase in voltage leads to increased power consumption. This demonstrates the interdependence between load and bus voltage: changes in load affect bus voltage, and conversely, variations in bus voltage impact the load. In other words, the load is dependent upon the bus voltage, and the bus voltage is dependent upon the load.

In this section, we will examine how this fact affects load flow.

Refer to Figure 3.10. When "Consider voltage dependency of loads" is taken into account, the voltage drop causes the loads to use less power, which is more in line with reality. Additionally, the 4.2 MW network losses have been decreased to 4.1 MW.

3.2.7 Reactive Power Compensation with Tap Changers

Transformer tap changers adjust the voltage ratio by slightly altering the number of turns in either the primary or secondary winding. The most affordable and useful voltage-correcting component in the network is a tap changer. Both under load (on load) and without load (off load) tap changers, accomplish this task.

Figures 3.11 to 3.13 are shown. The transformer tap changer is typically not required in normal mode. The bus-load voltage (No. 2) and, in turn, the transformer bus voltage (No. 3) drop when the load (No. 1) increases.

FIGURE 3.10
Analysis of the effect of "Consider voltage dependency of loads".

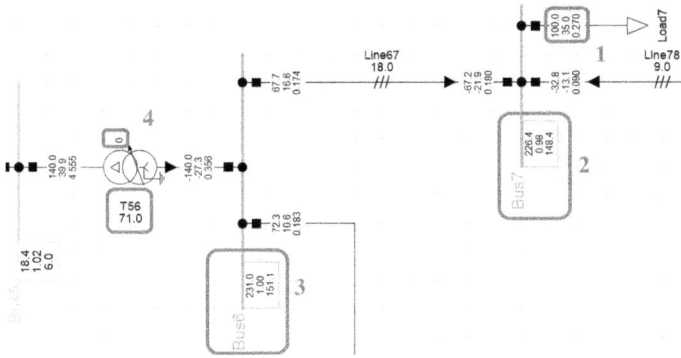

FIGURE 3.11
Tap changer not working in normal mode.

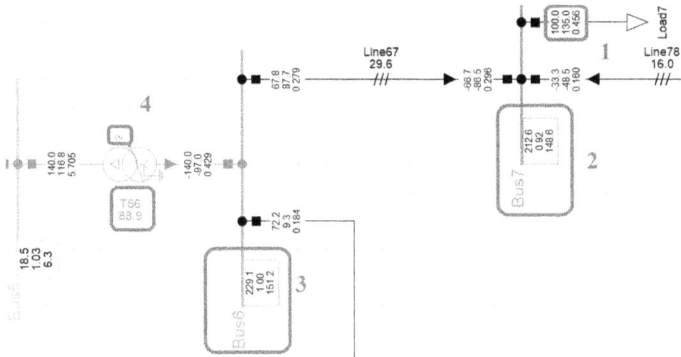

FIGURE 3.12
Tap changer operation in low load mode.

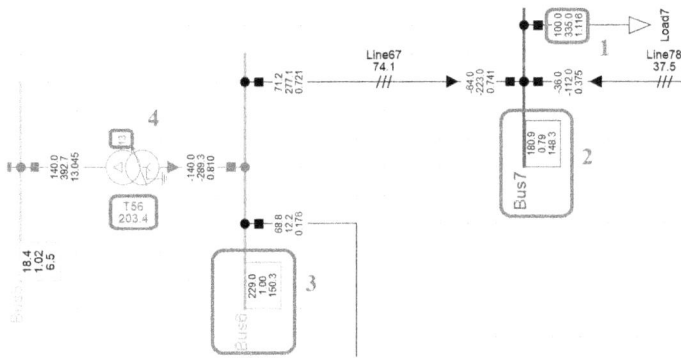

FIGURE 3.13
Tap changer operation in full load mode.

TABLE 3.2

Summary of Transformer Tap Performance

Tap	0	13	19	20	20	20	–
Voltage (pu)	1.00	1.00	0.99	0.98	0.96	0.88	No convergence in load flow!
Q Load (MVAr)	35	335	380	390	400	410	413

Therefore, the transformer tap operates (No. 4), and the transformer bus voltage (No. 3) and consequently the bus-load voltage (No. 2) increase. From low load to full load, this is carried out (Figure 3.13).

How long does the transformer tap run and what happens after that are the questions. The transformer tap is determined to answer this issue by gradually increasing the load until the network diverges.

Table 3.2 provides a summary of this investigation. The transformer tap rises in tandem with the load until it hits the maximum tap value, which in this case is 20. The voltage then drops as the transformer tap is fixed at its maximum. Until the network diverges, this keeps happening.

3.2.8 Reactive Power Compensation with Station Control

This kind of control raises a network point's voltage to the required level (often 1 per unit) by adjusting the generators' reactive power. Although technically possible, this approach is not useful in practice. Because the voltage cannot be remotely changed and is point-based. To demonstrate the idea, we attempt to manage generator G 5's reactive power in order to raise bus 7's voltage to one per unit. The impact of station control in regulating bus bar 7's voltage is depicted in Figure 3.14. The transformer and generator's voltage and loading are beyond the allowable range, even if bus bar 7's voltage has reached one per unit.

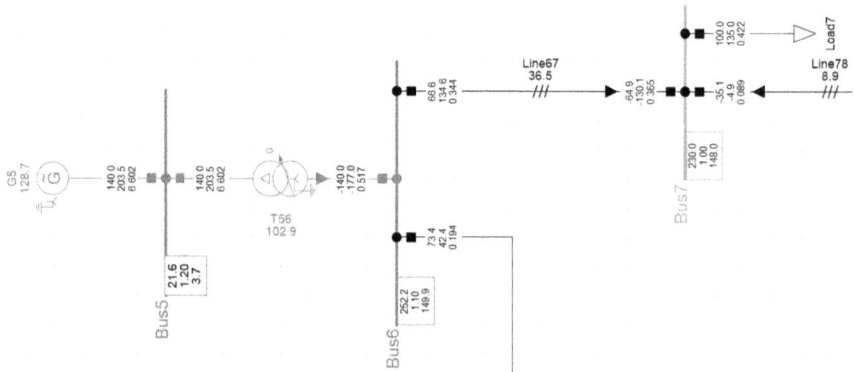

FIGURE 3.14
Station control effect.

3.2.9 Feeder Application

"Feeder" refers to "feeding point". A feeder is a location within the power network that supplies other components. Although feeders are more frequently utilized in radial distribution networks, we demonstrate two feeder uses here:

1. Changing loads with a known feeder power and
2. Drawing a feeder's voltage profile.

   1. The power at that location known as feeders typically has measured levels. The values specified for the load are probably incorrect if the power flowing through the feeder deviates from the actual value. Consequently, once the feeder's power is determined, the active and reactive loads are adjusted until the power flowing through the feeder is at the proper level. This is accomplished by altering the loads' scale.

      The network is shown in normal mode in Figure 3.15a. Instead of 54.9 MW, it is assumed that 100 MW of power is flowing through the feeder (Figure 3.15b). The active and reactive power of the loads can be altered by an "Adjusted Feeder by load Scaling" so that the feeder's power actually reaches 100 MW. Many distribution companies use this kind of analysis.

   2. The voltage at each feeder site is depicted in this figure as a function of distance from the feeder. This graph's benefit is that it allows one to identify any problematic place in the network by examining the feeder point.

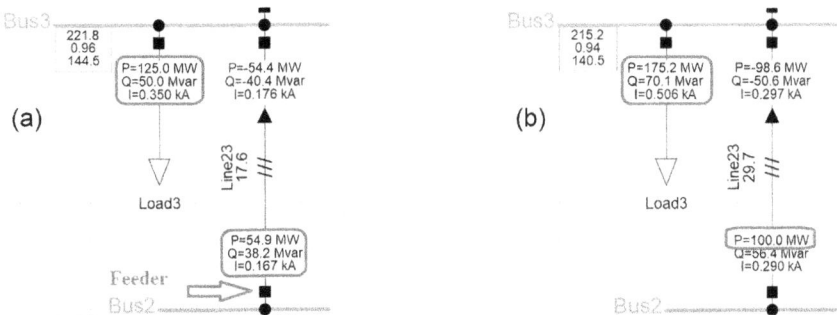

FIGURE 3.15
Effect of feeder power application on loads, (a) Normal mode, (b) Adjusted feeder by load scaling.

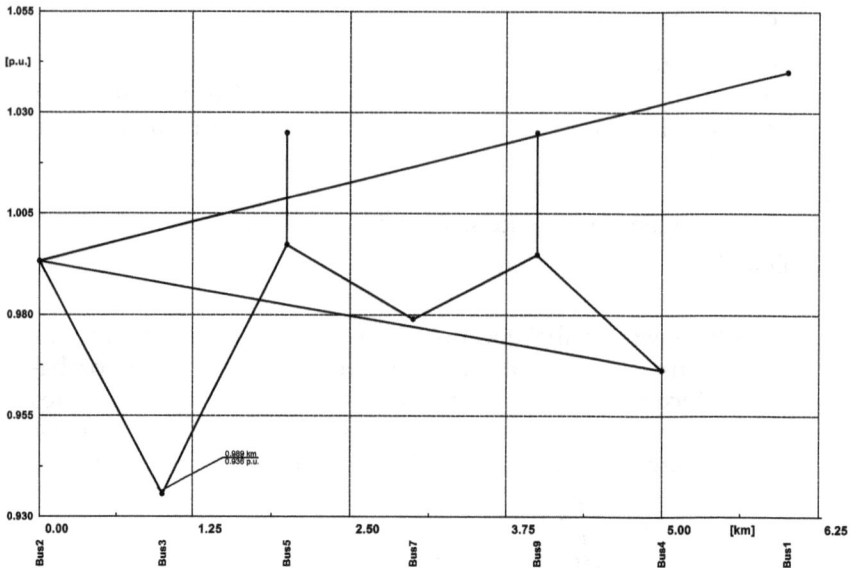

FIGURE 3.16
Voltage profile output.

An illustration of a voltage profile, or a plot of feeder voltage against distance from the feeder, is shown in Figure 3.16. The distance from the feeder point in this picture makes it easy to identify where the feeder's voltage is at its worst.

3.3 Two-Choice Questions (Yes/No)

1. Load flow analysis is a fundamental tool in power system engineering.
2. Load flow studies help in the planning and operation of power systems.
3. Voltage magnitude and phase angle are the primary variables in load flow analysis.
4. The Gauss–Seidel method is always faster than the Newton–Raphson method.
5. The Newton–Raphson method is widely used for load flow analysis due to its fast convergence.
6. Load flow analysis does not consider reactive power flow.

7. Fast-decoupled load flow is less accurate than the full Newton–Raphson method.

8. Load flow analysis is only concerned with the steady-state operation of a power system.

9. Power system control involves maintaining voltage and reactive power within specified limits.

10. Voltage control is not necessary in power systems.

11. Reactive power control is essential for maintaining voltage levels.

12. Tap-changing transformers can be used to control voltage levels.

13. Static VAR compensators (SVCs) only absorb reactive power.

14. Generator excitation systems do not affect voltage control.

15. Power system stabilizers are used to improve voltage stability.

16. Under-excitation limiters prevent generators from operating at low excitation levels.

17. Over-excitation limiters prevent generators from operating at high excitation levels.

18. Load tap-changing transformers are used to control voltage at the load side.

19. Load flow studies can help identify potential voltage violations.

20. Reactive power compensation can improve power system efficiency.

21. Optimal power flow is a technique for minimizing power system operating costs.

22. Reactive power compensation is only necessary during peak load periods.

23. OPF only considers economic factors.

24. The location of reactive power sources does not affect system performance.

25. OPF can be used to optimize power flow and voltage profiles.

26. The objective function in OPF is always linear.

27. The constraints in OPF can be equality or inequality constraints.

28. Load flow analysis can be used to assess the impact of renewable energy integration.

29. The solution to an OPF problem is always unique.

30. Energy storage systems can provide flexibility and support to power systems.

31. Demand response programs do not impact power system operation.

32. Load flow control is a static process.

3.3.1 Key Answers to Two-Choice Questions

Yes	1–3, 5, 7-9, 11, 12, 15-21, 25, 27, 28, 30
No	4, 6, 10, 13,–14, 22–24, 26, 29, 31, 32

3.4 Appendix, Load Flow Control Methods in DIgSILENT

This section requires a basic understanding of DIgSILENT PowerFactory software. You must download the file (Chapter3.pfd) from the book's end-of-book attachments in order to follow this section.

Step 1: Vector Plot

Run load flow. Right-click on any desired line and, show, and select the (Vector Diagram) or (Vector Plot), Current/Voltage option.

Step 2: Bar Diagram

First, run the load flow and then follow the steps below in order.
Select all busbars by holding down the Ctrl-A key
Right-click on a bus
Select Show
A Bar Diagram for voltages can be drawn in terms of per unit.

Step 3: Shunt capacitor

By choosing one of the icons (⚡), (⚡), (⏚), or (⚡) in this section, you can install an RLC filter shunt on each bus. You must adjust the parameters in (Design Parameter) and (Layout Parameter) to determine the available values of the RLC filter, such as reactive power, inductance, capacitance, etc., as seen in Figure 3.17.

Step 4: Automatic Tap changing

In this section, we will enable Automatic Tap changing of the T56 transformer. To do this, double-click on the T56 transformer. By default, you are in the **Basic Data** tab. Go to the **Load Flow** tab and enable the **Automatic Tap Changing** option. (As shown in Figure 3.18). Now, without changing the default numbers, go back to the **Basic Data** tab.

On the **Basic Data** tab, in the Type field, press the (➡) key. Then go to the **Load Flow** tab, then **Tap Changer**, and complete the information as shown in Figure 3.19. Now, press the OK key twice to return to the main page.

Then, run the Load Flow key (⚡) or (⚡) and activate the following option (Figure 3.20).

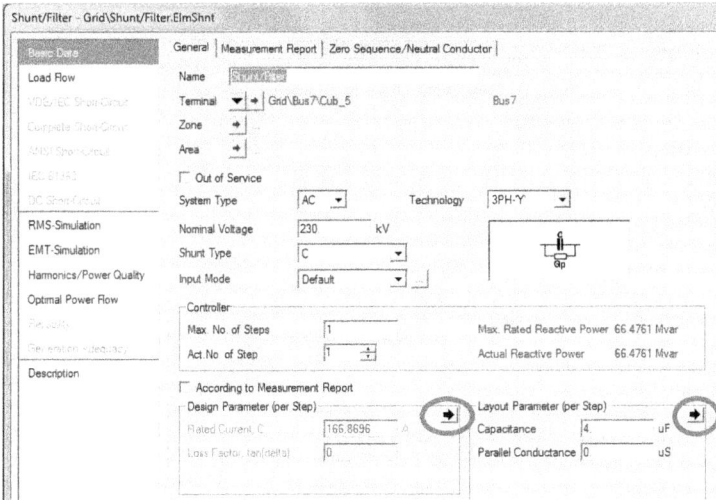

FIGURE 3.17
General information about an RLC shunt filter.

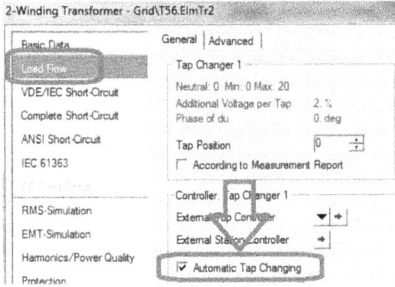

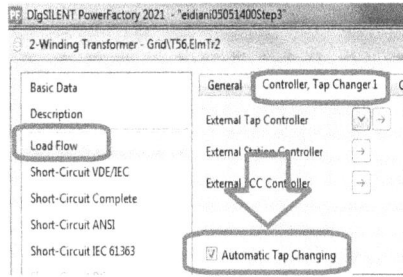

FIGURE 3.18
Activating the "Automatic Tap Changing" (v15.1 and v2021).

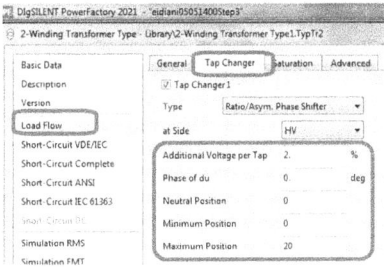

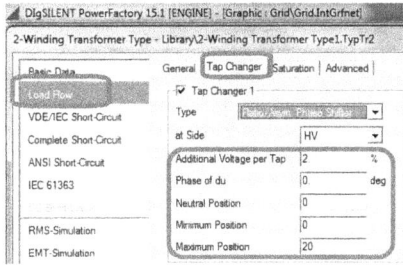

FIGURE 3.19
Input information to the transformer tap (v2021 and v15.1).

v15.1   v2021

Reactive Power Control ———————— Voltage and Reactive Power Regulation

☑ Automatic Tap Adjust of Transformers   ☑ Automatic tap adjustment of transformers

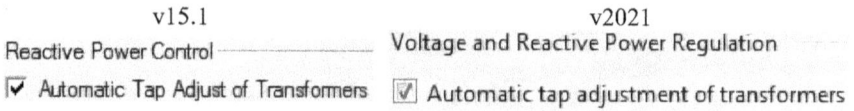

FIGURE 3.20
Transformer tap activation icon on the **Load Flow** tab.

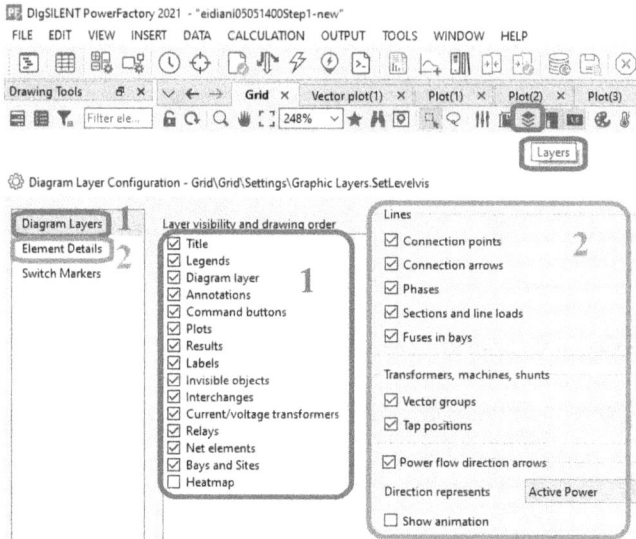

FIGURE 3.21
Activating different graphic layers.

Step 5: Layers

As seen in Figure 3.21, these layers must be engaged to view various layers, such as transformer tap changers, power direction, etc.

Step 6: Station control

On the **Load Flow** tab, as illustrated in Figure 3.22, right-click on one of the generators, choose **Define**, then **Station control**, and click

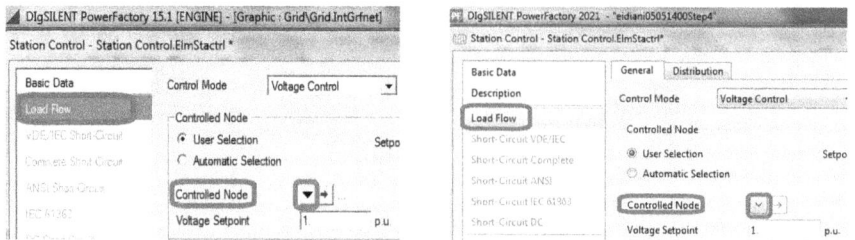

FIGURE 3.22
Generator-controlled bus selection.

the button on the other side **Controlled Bar** or **Controlled Node**. Next, press the **"Select"** button. Choose the bus whose voltage has to be regulated in the active study case. Run the load flow after selecting OK. In this instance, the generator regulates the bus voltage.

Step 7: Feeder

See Figure 3.23. To define a feeder, you must right-click on the feeder key, click **Define** and **Feeder**, and in the **Load Flow** section, set the **Active Power** to 50 MW. It is necessary to verify the **"Adjusted by Feeder Load Scaling"** of the intended loads, as seen in Figure 3.24. Lastly, the Load flow tab's (☑ Feeder Load Scaling) option needs to be enabled.

Next, you can choose **Show** by right-clicking on one of the lines (optional), followed by **Voltage Profile**. We'll sketch the intended feeder curve.

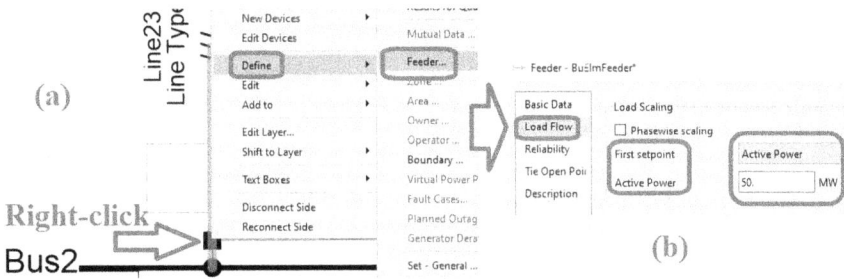

FIGURE 3.23
Feeder definition.

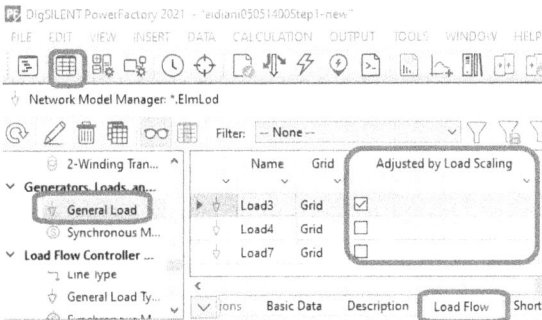

FIGURE 3.24
Adjusted by Feeder Load Scaling.

3.5 Summary

This chapter focuses on using DIgSILENT PowerFactory software to teach load flow analysis, a critical aspect of power system engineering that determines steady-state operational parameters like voltage levels and power flows. Load flow analysis is essential for planning and managing power systems. Various control strategies, including voltage and reactive power management, are discussed, along with methods for solving power flow equations, such as Newton–Raphson and Gauss–Seidel. This chapter emphasizes the importance of these techniques in ensuring reliable and efficient grid operation, with additional guidance provided in the appendix for further exploration.

Note

1 The term, "load flow" and "power flow" are the same thing—they are interchangeable terms used in power system engineering. Both refer to the same analysis that determines how power flows through an electrical network under steady-state conditions.

  The term "load flow" emerged historically because the analysis focuses on determining how power flows to meet the various loads (power demands) in the system. "Power flow" became more commonly used later as it directly describes what's being analyzed—the flow of power (both active and reactive) through the network elements. Neither term is more technically correct than the other. In industry practice, "load flow" remains the more commonly used term among power system engineers.

References

1. Eidiani, M., "An efficient differential equation load flow method to assess dynamic available transfer capability with wind farms," *IET Renewable Power Generation*, 2021, 15, pp. 3843–3855, https://doi.org/10.1049/rpg2.12299.
2. Eidiani, M., "A new load flow method to assess the static available transfer capability," *Journal of Electrical Engineering and Technology*, 2022, 17(5), pp. 2693–2701, https://doi.org/10.1007/s42835-022-01105-3.
3. Eidiani, M., Ashkhane, Y., Khederzadeh, M., "Reactive power compensation in order to improve static voltage stability in a network with wind generation," *2009 International Conference on Computer and Electrical Engineering, ICCEE 2009*, 2009, vol. 1, pp. 47–50, https://doi.org/10.1109/ICCEE.2009.239.

4. Eidiani, M., Zeynal, H., Zadeh, A.K., Mansoorzadeh, S., Nor, K.M., "Voltage stability assessment: An approach with expanded Newton Raphson-Sydel," *2011 5th International Power Engineering and Optimization Conference*, 2011, pp. 31–35, https://doi.org/10.1109/PEOCO.2011.5970424.
5. Eidiani, M., "A reliable and efficient method for assessing voltage stability in transmission and distribution networks," *International Journal of Electrical Power and Energy Systems*, 2011, 33(3), pp. 453–456, https://doi.org/10.1016/j.ijepes.2010.10.007.
6. Eidiani, M., "A new method for assessment of voltage stability in transmission and distribution networks," *International Review of Electrical Engineering*, 2010, 5(1), pp. 234–240.
7. Eidiani, M., "Assessment of voltage stability with new NRS," *2008 IEEE 2nd International Power and Energy Conference*, 2008, pp. 494–496, https://doi.org/10.1109/PECON.2008.4762525.
8. Eidiani, M., Buygi, M.O., Ahmadi, S., "CTV, complex transient and voltage stability: A new method for computing dynamic ATC," *International Journal of Power and Energy Systems*, 2006, 26(3), pp. 296–304, https://doi.org/10.2316/Journal.203.2006.3.203-3597.
9. Eidiani, M., Kargar, M., "Frequency and voltage stability of the microgrid with the penetration of renewable sources," *2022 9th Iranian Conference on Renewable Energy & Distributed Generation (ICREDG)*, 2022, pp. 1–6, https://doi.org/10.1109/ICREDG54199.2022.9804542.

4

Short-Circuit Analysis Techniques

4.1 Introduction

This chapter uses DIgSILENT PowerFactory software to teach short-circuit analysis and the necessary information. As you are aware, setting the relay and figuring out the switch-breaking capacity requires the short-circuit current value. In PowerFactory or other professional software, short-circuit analysis involves a number of settings and options that can yield varying outcomes. It should be possible for users to modify these settings to customize the results to suit their own requirements.

Important considerations include determining the maximum branch current, taking protective devices into account, and selecting sub- or transient values for the relay breaking time. Users can also choose phase matrices for unbalanced short-circuit calculations, disregard transient calculations, or take the motor contribution into account when calculating minimum short circuits. The different aspects of short-circuit analysis in DIgSILENT PowerFactory are covered in this chapter.

How to use the software and the possibility for further work are explained in the appendix. For more detailed work and further work, you can refer to the references [1–14] at the end of this chapter.

4.2 An Overview of Short-Circuit Analysis

Electrical systems must be designed and operated with short-circuit analysis in order to safely and effectively handle unforeseen short-circuit conditions. Significant problems, such as equipment damage and worker safety hazards, can result from short circuits. Systems must be built to isolate faults with the least amount of disturbance in order to reduce these risks. Environmental variables, insulation deterioration, and equipment failure are some of the causes of short circuits. To determine the maximum and minimum expected currents during the planning phase, which inform the design of protection schemes and equipment sizing, short-circuit calculations are crucial.

 DOI: 10.1201/9781003590514-4

There are various techniques for calculating short circuits, and each is appropriate for a particular use case. For example, IEC 61660 and ANSI/ IEEE 946 are designed for DC systems, whereas IEC 60909 and ANSI (*e.g.,* ANSI C37 standards) are frequently used for AC short-circuit calculations. The Complete Method is especially helpful for established systems with comprehensive operational data because it provides a more accurate analysis by taking into account the network's actual operating conditions. On the other hand, simplified techniques such as IEC60909, which offer conservative safety estimates, are useful during the design stage when comprehensive data may not be available.

With the aid of PowerFactory software, users can simulate single and multiple faults with high complexity using a variety of short-circuit calculation techniques. The software allows for the analysis of the effects of asynchronous motors and protection devices, as well as options for various fault types and impedance considerations. Through the toolbar or context-sensitive menus, users can perform calculations and have the option to create comprehensive reports on the outcomes. Engineers can efficiently evaluate and handle short-circuit situations in electrical networks thanks to this flexibility in analysis and reporting.

This chapter covers DIgSILENT PowerFactory's short-circuit calculation features and options, such as mixed mode evaluations, basic and advanced options for various standards, and the creation of reports and results. It describes how users can set up pre-load conditions, choose calculation methods, and examine results using a variety of visualization tools, such as tabular reports and single-line diagrams.

4.3 Short-Circuit Analysis

The **Methods** and **Fault Types** that can be checked in the DIgSILENT software are examined first. This information is displayed in Figure 4.1.

Consider the same nine-bus network of Chapter 3. A three-phase symmetrical short circuit in accordance with the IEC 60909 standard is examined first. Once the short-circuit current has been calculated, two output types can be taken into consideration. See Figure 4.2. The currents are represented graphically in one and numerically in the other. According to Figure 4.2, buses 1, 9, and 5 appear to be the worst in terms of short circuits (Ikss). It is preferable to use power rather than current when comparing buses, though. The short-circuit power (Skss) is shown in Figure 4.3. The worst buses in this figure in terms of short-circuit power are buses 1, 5, and 2. Note that the impact of each bus is calculated separately.

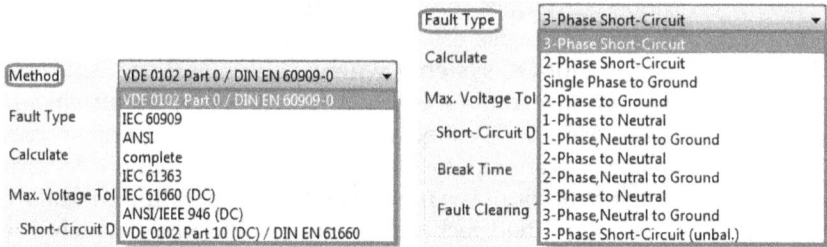

FIGURE 4.1
The methods and fault types in the DIgSILENT.

4.3.1 Comparison of Different Short Circuits

This section uses different standards to compare the three-phase and single-phase short-circuit currents on two buses, the generator (Bus 9) and the load bus (Bus 7). See Tables 4.1 and 4.2. The tables provide a few conclusions.

1. The short-circuit currents are the same in both standards VDE and IEC60909.
2. The minimum and maximum short-circuit currents are not very different.
3. The complete short-circuit current is between the maximum and minimum short-circuit currents of the standards.
4. On the generator bus, the single-phase short-circuit current is usually greater than the three-phase short-circuit current.
5. On the load bus, the three-phase short-circuit current is usually greater than the single-phase short-circuit current.

4.3.2 Multiple Fault

Another type of short circuit is a simultaneous short circuit which is defined as a multiple fault. The three-phase short-circuit current of buses 6 and 8 is first provided independently in Table 4.3. Additionally displayed is the three-phase simultaneous short-circuit current of buses 6 and 8. It is evident that the simultaneous short-circuit current poses less risk than the independent connection.

4.3.3 Short Circuit at Line

Table 4.4, Figure 4.4, and Figure 4.5 show the standard IEC 60909, three-phase short-circuit current on buses 3 and 6 as well as the short-circuit on lines 3–6 in terms of various line short-circuit percentages.

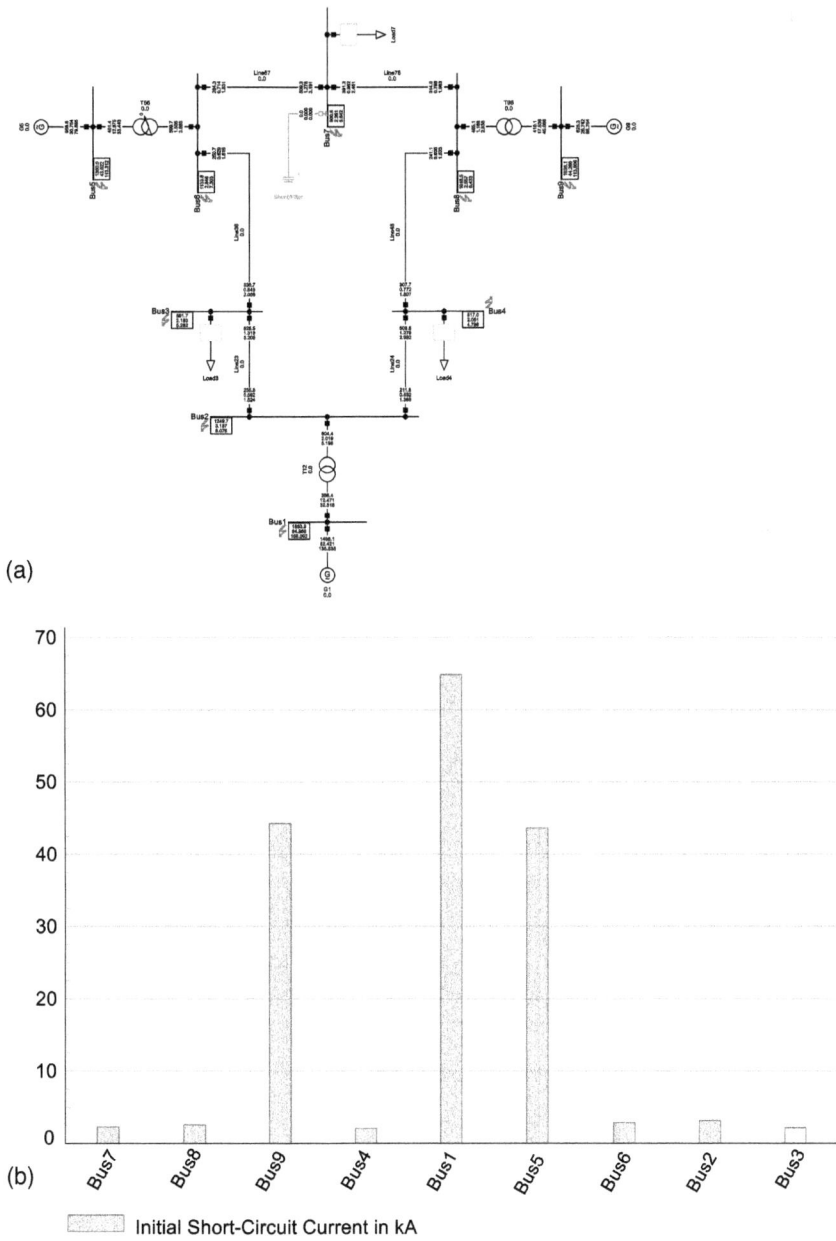

(a)

(b)

Initial Short-Circuit Current in kA

FIGURE 4.2
Software short-circuit current outputs (a) graphically, (b) numerically.

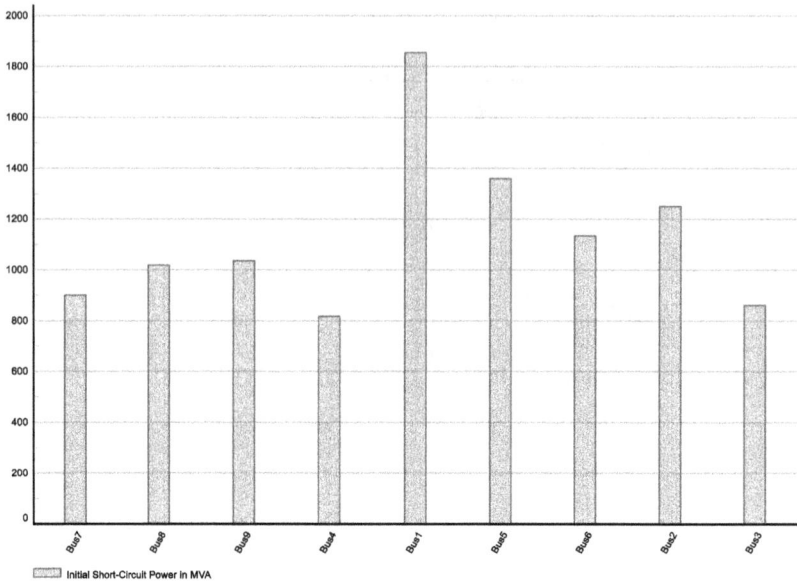

Initial Short-Circuit Power in MVA

FIGURE 4.3
Software short-circuit power outputs.

TABLE 4.1

Comparison of Different Short Circuits (Ikss in Bus9)-Generator Bus

Standard	VDE	IEC 60909	ANSI	Complete	VDE	IEC 60909	ANSI	Complete
Phase		3ph				1ph		
Max	50.351	50.351	43.351	46.934	53.015	53.015	46.003	48.808
Min	44.266	44.266		46.889	47.005	47.005		48.788

TABLE 4.2

Comparison of Different Short Circuits (Ikss in Bus7)-Load Bus

Standard	VDE	IEC 60909	ANSI	Complete	VDE	IEC 60909	ANSI	Complete
Phase		3 ph				1 ph		
Max	2.855	2.855	2.265	2.568	2.541	2.541	2.115	2.402
Min	2.261	2.261		2.564	2.112	2.112		2.399

The short-circuit current on the line is not a linear value; rather, it changes in the shape of a curve, as shown in Figure 4.5.

TABLE 4.3

Comparison of Three-Phase Simultaneous
and Multiple Fault Short Circuits

Separately Short Circuits (Independently)		Multiple Fault Short Circuits (Simultaneous)	
Bus 6	Bus 8	Bus 6	Bus 8
3.204	2.765	2.391	1.825

TABLE 4.4

Three-Phase Short-Circuit Current on Buses 3 and 6 and Line 3–6 Fault Distance
from Bus 6 in Percent

Location %	Bus 6 (0%)	10	20	30	40	50	60	70	80	90	Bus 3 (100%)
Current	4.22	3.644	3.253	2.979	2.78	2.65	2.55	2.49	2.466	2.469	2.50

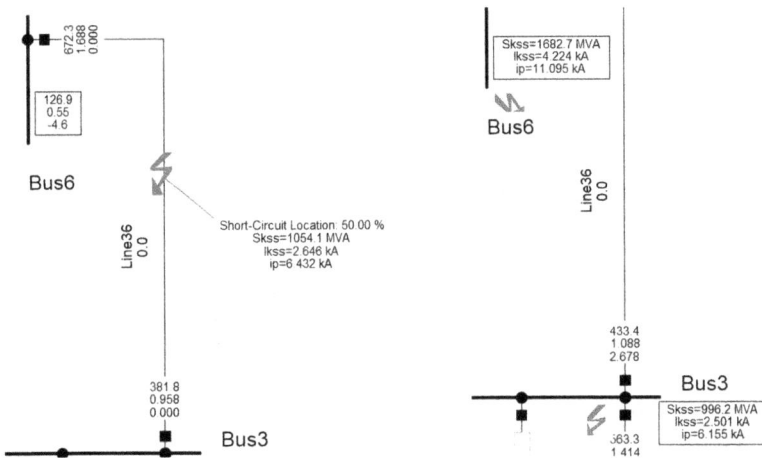

FIGURE 4.4
Short-circuit current on bus and line.

4.4 Two-Choice Questions (Yes/No)

1. A short circuit is an abnormal condition in a power system where an unintended connection occurs between two points of different potential.

2. Symmetrical faults are more common than asymmetrical faults in power systems.

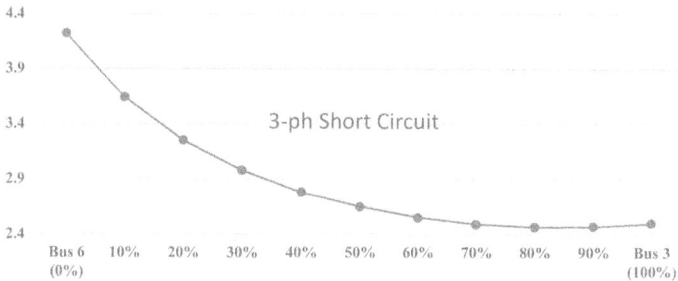

FIGURE 4.5
Short-circuit current curve on the line between bus 6 and bus 3.

3. Asymmetrical faults do not cause significant disturbances in power systems.

4. Line-to-line faults do not cause significant voltage unbalance.

5. Double-line-to-ground faults can lead to high fault currents.

6. A symmetrical fault involves a short circuit between all three phases of a three-phase system.

7. Single-line-to-ground faults are the most common type of asymmetrical fault.

8. Symmetrical faults are easier to analyze than asymmetrical faults.

9. The zero-sequence component of current is zero in a symmetrical fault.

10. The negative-sequence component of current is zero in a single-line-to-ground fault.

11. Symmetrical components are used to analyze only balanced three-phase systems.

12. The symmetrical component transformation is a mathematical technique used to simplify the analysis of unbalanced systems.

13. The positive-sequence component of voltage is zero in a line-to-line fault.

14. The zero-sequence network is not affected by the grounding impedance of the system.

15. The negative-sequence network is a mirror image of the positive-sequence network.

16. The zero-sequence network represents the unbalanced component of the system.

17. The positive-sequence network represents the balanced component of the system.

18. The fault current in an asymmetrical fault can be significantly higher than the fault current in a symmetrical fault.

19. The transient component of the fault current decays very slowly.

20. The fault current in a symmetrical fault is always lower than the fault current in an asymmetrical fault.

21. The steady-state component of the fault current is determined by the system impedance.

22. The transient component of the fault current is not affected by system parameters.

23. Circuit breakers are not used to interrupt fault currents.

24. Relays are used to detect and isolate faults.

25. The fault current can cause excessive heating and damage to equipment.

26. Overcurrent relays are only used to detect overcurrent faults.

27. Differential relays are used to protect transformers.

28. Ground fault relays are used to protect against ground faults.

29. Distance relays can measure the distance to a fault.

30. Proper coordination of protective devices can minimize the impact of faults.

31. The selection of protective devices is not influenced by fault currents.

32. The coordination of protective devices is not important in power systems.

33. The performance of protective devices is not affected by system disturbances.

34. The reliability of power systems is improved by effective fault protection.

35. The setting of protective devices is critical to ensure proper operation.

4.4.1 Key Answers to Two-Choice Questions

Yes	1, 5–8, 9, 12, 15–18, 21, 24, 25, 27–30, 34, 35
No	2–4, 10, 11, 13–14, 19, 20, 22, 23, 26, 31–33

4.5 Appendix, Short-Circuit Analysis in DIgSILENT

This section requires a basic understanding of DIgSILENT PowerFactory software. You must download the file (Chapter4.pfd) from the book's end-of-book attachments in order to follow this section.

Step 1: Short-Circuit Calculations

Short-circuit calculations can be carried out with standard types on all buses or at a particular location, as illustrated in Figure 4.6.

Step 2: Multiple Fault Short-Circuit Calculations

See Figure 4.7.

1. and 2. Select the buses Bus6 and Bus8 by holding the Control key (Ctrl).

3. Then right-click and select the **Calculate** option.

4. Then run the **Multiple Faults** option and press the Close button.

5. By pressing the Execute button, the software performs a simultaneous short circuit on the two buses using the Complete Method.

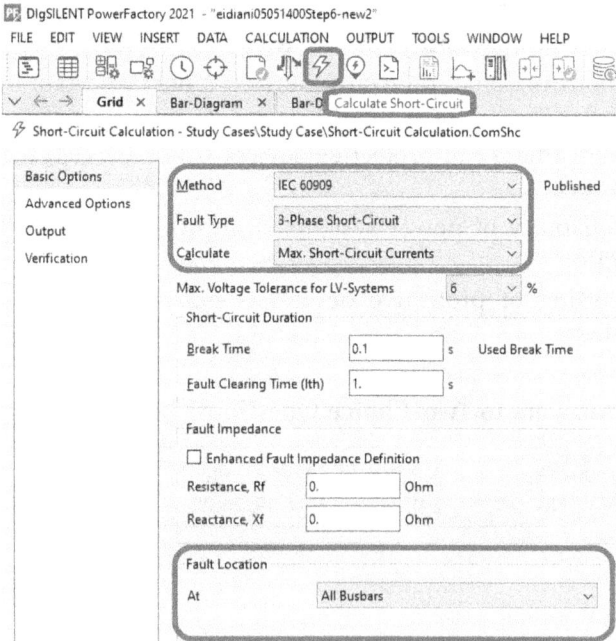

FIGURE 4.6
Calculate the short-circuit window.

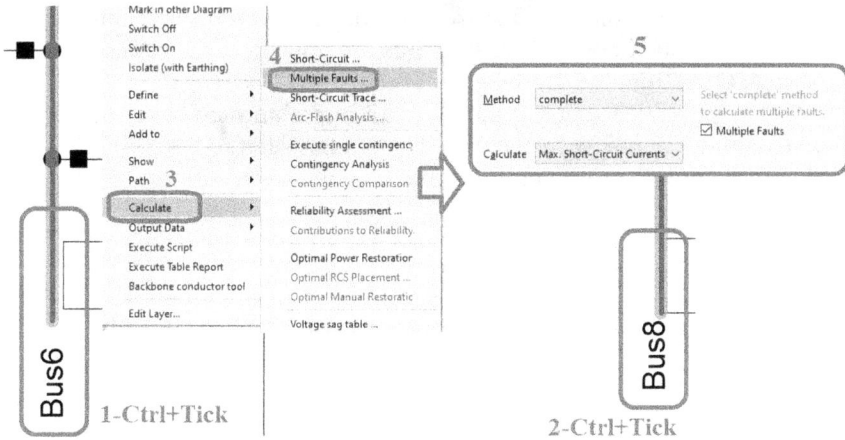

FIGURE 4.7
Multiple fault short-circuit calculation.

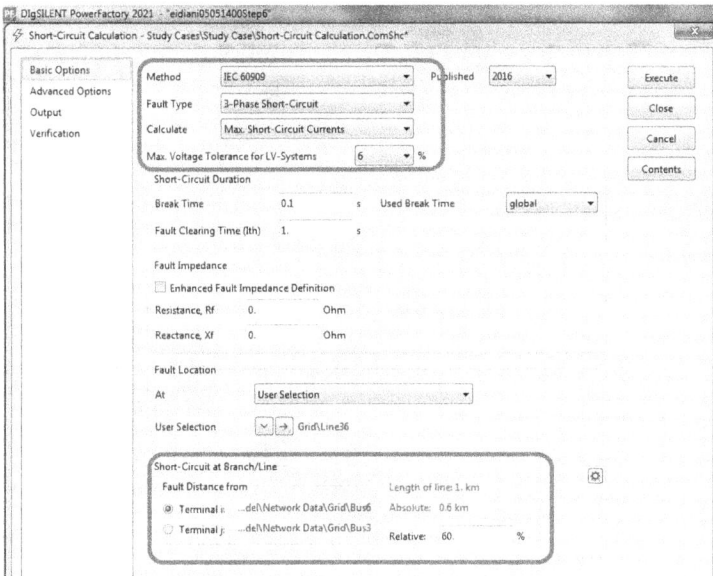

FIGURE 4.8
Short-circuit information on the line.

Step 3: Calculate Short Circuit at the Branch/Line

In the **Calculate** option, right-click on a line (for instance, Line36) and choose the **Short-Circuit option**. As seen in Figure 4.8, a figure that enables you to execute a short circuit using the IEC method at a ratio of 60% of Bus6 appears. The percentage of fault is modifiable. You can choose this percentage range from 0% to

100%, as illustrated in Figure 4.5. The short circuit is located on Bus 6 if you choose 0%, and on Bus 3 if you choose 100%.

Step 4: Question

Is it possible to set the short-circuit impedance on buses 6 and 8 to 2 and 1 Ω, respectively? Yes. The reader is left to figure out how to accomplish this, even though it is very easy.

4.6 Summary

This chapter focuses on using DIgSILENT PowerFactory software for short-circuit analysis, emphasizing the importance of short-circuit current in relay settings and switch-breaking capacity. It highlights the customizable nature of the software, allowing users to adjust various settings for tailored results, such as maximum branch current, protective devices, and relay breaking times. Additionally, it covers options for unbalanced short-circuit calculations and motor contributions. The appendix guides software usage and suggests further reading for in-depth exploration.

References

1. Kasikci, I., *Short Circuits in Power Systems: A Practical Guide to IEC 60909-0*, Wiley-VCH, 2nd edition, 2018.
2. Eidiani, M., Heidari, V., *Fundamentals of Power Systems Analysis 1: Problems and Solutions*, Taylor & Francis Group, CRC Press, 2023, pp. 1–215, https://doi.org/10.1201/9781003394433
3. Eidiani, M., Rouzbehi, K., *Advanced Topics in Power Systems Analysis: Problems, Methods, and Solutions*, Taylor & Francis Group, CRC Press, 2024, pp. 1–120. http://www.routledge.com/9781032828664
4. Eidiani, M., "Modeling renewable energy resources using DIgSILENT PowerFactory software," In: Chenniappan, S., Padmanaban, S., Palanisamy, S. (eds) *Power Systems Operation with 100% Renewable Energy Sources*, Elsevier, 2024, pp. 165–202. https://doi.org/10.1016/B978-0-443-15578-9.00013-3
5. Eidiani, M., "The effect of power system strength on the calculation of available transmission capacity," In: Alhelou, H.H., Hosseinzadeh, N., Bahrani, B. (eds) *Power System Strength: Evaluation Methods, Best Practice, Case Studies, and Applications*, 2024, pp. 137–174, https://doi.org/10.1049/PBPO247E_ch7
6. Mohammadi, F., Rouzbehi, K., Hajian, M., Niayesh, K., Gharehpetian, G. B., "HVDC circuit breakers: A comprehensive review," *IEEE Transactions on Power Electronics*, vol. 36, no. 12, pp. 13726–13739, 2021.

7. Heidary, A., Radmanesh, H., Bakhshi, A., Rouzbehi, K., Pouresmaeil, E., "A compound current limiter and circuit breaker," *Electronics*, vol. 8, no. 5, pp. 551, 2019.

8. Heidary, A., Radmanesh, H., Moghim, A., Ghorbanyan, K., Rouzbehi, K., "A multi-inductor H bridge fault current limiter," *Electronics*, vol. 8, no. 7, pp. 795, 2019.

9. Heidary, A., Rouzbehi, K., Mehrizi-Sani, A., Sood, V. K., "A self-activated fault current limiter for distribution network protection," *IEEE Journal of Emerging and Selected Topics in Power Electronics*, vol. 10, no. 4, 2021.

10. Heidary, A., Radmanesh, H., Bakhshi, A., Samandarpour, S., Rouzbehi, K., "Compound ferroresonance overvoltage and fault current limiter for power system protection," *IET Energy Systems Integration*, vol. 2, no. 4, pp. 325–330, 2020.

11. Hesami, M., Rouzbehi, K., Moradlou, M., Mousavi, S. M., Behzadpoor, S., "Fault Current Protection in distribution system connected EVCH: A Review," *2021 25th Electrical Power Distribution Conference* (EPDC), pp. 76–83, 2021.

12. Heidary, A., Hesami, M., Bakhshi, A., Rouzbehi, K., "Wind energy generators fault Current protection: Structures survey," *7th Iran Wind Energy Conference* (IWEC2021), pp. 1–6, 2021.

13. Heidary, A., Radmanesh, H., Naghibi, S.H., Samandarpour, S., Rouzbehi, K., "Distribution system protection by coordinated fault current limiters," *IET Energy Systems Integration* 2 (1), 59–65, 2020.

14. Hesami, M., Bigdeli, M., Rouzbehi, K., Firouzi, M., Shafaghatian, N., "A Dual Function Protection System Based on Fault Current Limiter and Circuit Breaker for DFIG," *2023 3rd International Conference on Electrical Machines and Drives* (ICEMD), pp. 1–7, 2023.

5

Nonlinear Dynamic Analysis Methods

5.1 Introduction

This chapter uses DIgSILENT PowerFactory software to teach transient stability analysis and the necessary information. Analysis of Nonlinear Dynamics Methods are computational approaches that consider nonlinear components such as generators, loads, and control systems to investigate how power systems behave under dynamic circumstances. These techniques are essential for comprehending and forecasting how the system will react to disruptions like faults, load changes, or abrupt generation fluctuations.

One particular use of nonlinear dynamic analysis is transient stability analysis, which evaluates the power system's capacity to preserve synchronism and voltage stability after a disturbance. The ability of the system to recover from the disturbance and return to a stable operating state can be assessed by engineers by simulating its dynamic behavior over a brief time horizon.

More than 2000 parameters can be used to create dynamic curves with DIgSILENT software. In this chapter, we will discuss one example of them. The goal of this section is to ascertain the critical clearing time.

How to use the software and the possibility for further work are explained in the appendix. For more detailed work and further work, you can refer to the references [1–20] at the end of this chapter.

5.2 Simulation RMS/EMT

In power system analysis, two distinct simulation techniques are employed: EMT (Electromagnetic Transient) and RMS (Root Mean Square). EMT simulations offer a thorough time-domain examination of power systems, capturing fast transient phenomena like switching operations, faults, lightning strikes, and so on. To model the behavior of different components, such as generators, transformers, and transmission

DOI: 10.1201/9781003590514-5

lines, differential equations must be solved. Despite their high accuracy, EMT simulations require a lot of computing power.

In contrast, RMS simulations analyze power systems in the frequency domain using phasor representations. They work well for researching slower dynamic phenomena like stability analysis and power flow. RMS simulations normally require less computing power than EMT simulations, and they do not offer as much detailed information about fast transients.

This chapter examines the RMS simulation approach used to calculate the critical clearing time. You can see the references [1–22] at the end of this chapter for more in-depth work and future research.

5.3 Transient Stability Analysis (TSA)

Consider the 9-bus system from earlier chapters. The appendix at the end of this chapter contains instructions on how to run these brief simulations. Let's say the network has experienced a major fault. A short circuit, a line break, or any other type of fault could be the cause. Assume, for instance, that a three-phase short circuit on Line36 happens in the middle of the line at 0.1 second, and that the line switches open to clear the short circuit at 0.3 second.

In DIgSILENT PowerFactory, rotor angles play a critical role in analyzing the dynamic behavior and stability of synchronous machines within a power system. Several key terms describe different aspects of rotor angles and their relationships to the system's operation.

The **fipol**, or internal rotor angle, represents the angular displacement between the rotor's magnetic field and the stator's magnetic field. This angle, also known as the load angle among industry experts, is vital for understanding the synchronization and stability of a generator in the power system.

The **firel**, or relative rotor angle, takes this concept further by comparing the rotor angle of a generator to that of the system's reference machine, often the slack or swing generator. This relative measurement helps assess how individual generators align dynamically with the system's overall performance.

Another crucial parameter is the **firot**, which measures the rotor angle with respect to the reference voltage of the network. This reference voltage is typically provided by the slack bus, serving as the system's anchor point. By relating the rotor angle to this reference, firot provides insights into the generator's phase relationship with the broader network.

Finally, **dfrotx** quantifies the maximum deviation of rotor angles among all synchronous machines in the system. This value highlights the largest angular difference between machines, which is a critical indicator of system stability. Significant deviations in dfrotx can suggest potential instability or risk of losing synchronism between generators, especially during disturbances or transient events.

Together, these parameters provide a comprehensive view of rotor dynamics and stability, helping engineers ensure the reliable operation of power systems.

In DIgSILENT PowerFactory, the angle **fi** provides critical insight into the position of the rotor in relation to the network's reference voltage. It is expressed in two different units for flexibility and precision: degrees (**fi [deg]**) and radians (**phi [rad]**).

When expressed in degrees, **fi [deg]** represents the angular position of the rotor's direct axis (**d-axis**) relative to the network's reference voltage. This format is commonly used for intuitive interpretation in studies and presentations.

The same angle, when expressed in radians as **phi [rad]**, retains the same physical meaning: it describes the position of the rotor's **d-axis** relative to the reference voltage. The use of radians is particularly useful in mathematical and computational analyses due to its direct compatibility with trigonometric functions.

These parameters are indispensable in understanding the machine's synchronization and stability within the power system, as they reflect how the rotor aligns with the broader network dynamics (Figure 5.1).

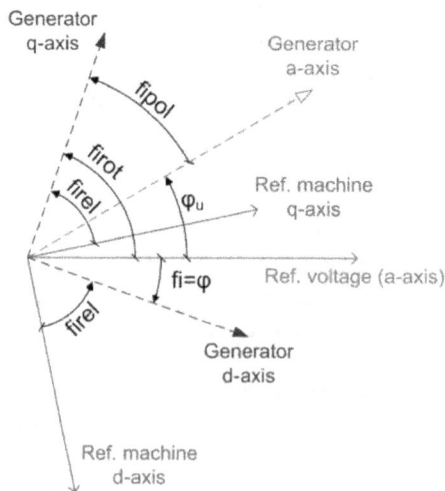

FIGURE 5.1
Rotor angles defined by DIgSILENT.

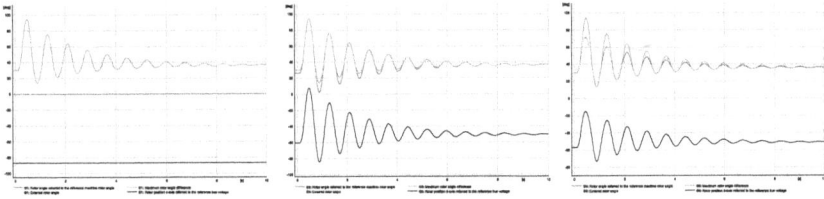

FIGURE 5.2
The output of three generators at different angles in the steady-state network, $T_{cl}=0.2$ (0.3−0.1.)

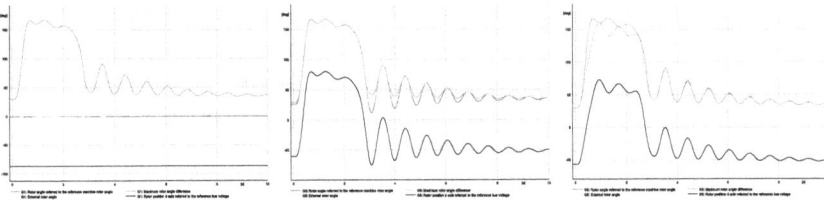

FIGURE 5.3
The output of three generators at different angles in the stability boundary network, $T_{cl}=0.330005$ (0.430005−0.1.)

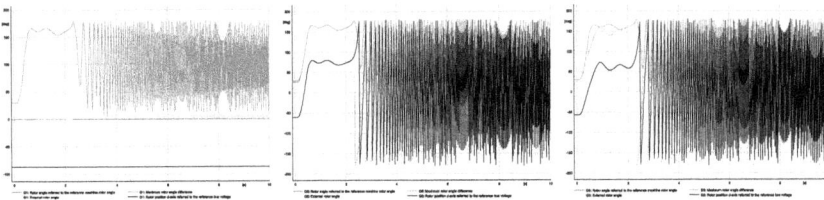

FIGURE 5.4
The output of three generators at different angles in an unstable state network, $T_{cl}=0.3300051$ (0.4300051−0.1.)

With the short-circuit fault above, the different angles are plotted in Figure 5.2. The fault clearing time (T_{cl}) in this case is 0.2 second (=0.3−0.1). Keep in mind that the network's stability and oscillation reduction are the only things that matter, regardless of the generator angle definition that is applied. In Figure 5.3, the line clearing time is $T_{cl}=0.330005$ (=0.430005−0.1) and in Figure 5.4, the fault clearing time is $T_{cl}=0.3300051$ (=0.4300051−0.1). Notice that nothing happened for 0.1 second.

Figures 5.2 to 5.4 show that the critical clearing time is 0.330005 second, meaning that the network becomes unstable when the fault clearing time is added to 1e-7. It is clear that only one of the generator definitions can be applied, and the critical clearing angle calculation is unaffected by the type of angle definition.

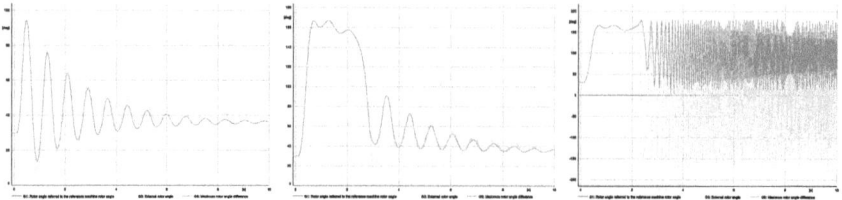

FIGURE 5.5
The output of three generators with **"firel"** angle in three different states.

The "firel" angle definition, representing the relative rotor angle of each generator with respect to the reference generator, is one of the most widely recognized and utilized concepts in power system analysis. As a result, during the simulation, the reference generator "G1" of this type is zero. In this instance, the critical clearing time is constant, as shown in Figure 5.5.

It is important to note that in all unstable transient network states, generator angles tend to diverge to infinity. However, in DIgSILENT PowerFactory, the software rewraps the angles back to –180° once they exceed 180°. As a result, in unstable conditions, the angles appear jagged due to this periodic wrapping.

5.4 Two-Choice Questions (Yes/No)

1. Transient stability analysis is a critical tool for assessing the dynamic behavior of power systems.
2. A power system is considered transiently stable if it can maintain synchronism after a disturbance.
3. Small disturbances, like load changes, do not affect transient stability.
4. Large disturbances, such as short circuits, can lead to transient instability.
5. Generator inertia does not play a role in transient stability.
6. Transient stability is concerned with the long-term stability of a power system.
7. The strength of the power system, as measured by the transfer reactance, affects transient stability.
8. Power system damping can enhance transient stability.
9. Excitation systems do not affect transient stability.
10. Increased loading always improves transient stability.

11. Transient stability analysis is only concerned with the rotor angle stability of generators.
12. Governor systems can influence transient stability.
13. Voltage stability is also a concern in transient stability analysis.
14. Transient stability analysis is not affected by system configuration changes.
15. Power system simulations are used to assess transient stability.
16. The time-domain simulation method is accurate but computationally intensive.
17. Time-domain simulations are not suitable for transient stability analysis.
18. The eigenvalue method is not used for transient stability analysis.
19. The equal-area criterion is a simplified method for assessing transient stability.
20. The critical clearing time is a key parameter in transient stability analysis.
21. Increasing the generator inertia can decrease the critical clearing time.
22. The equal-area criterion is always accurate.
23. Flexible AC Transmission Systems (FACTS) devices do not affect transient stability.
24. Transient stability analysis is only applicable to large power systems.
25. Power system stabilizers can improve transient stability.
26. High-voltage direct current (HVDC) transmission can influence the transient stability.
27. Small-scale power systems can also experience transient stability issues.
28. Renewable energy sources do not impact transient stability.
29. The integration of renewable energy sources can pose challenges to transient stability.
30. Transient stability analysis is a type of dynamic analysis.
31. The initial conditions of the power system do not affect transient stability.
32. Transient stability analysis is a static analysis.
33. Transient stability is only concerned with the behavior of generators.
34. The initial conditions of the power system can significantly impact transient stability.
35. Transient stability is concerned with the behavior of the entire power system.

36. The impact of disturbances on transient stability can vary depending on system conditions.
37. Transient stability analysis is a one-time process.
38. The impact of disturbances on transient stability is always the same.
39. Transient stability analysis is an ongoing process, especially as power systems evolve.
40. Transient stability is not a concern in modern power systems.

5.4.1 Key Answers to Two-Choice Questions

Yes	1, 2, 4, 7, 8, 12, 13, 15, 16, 19, 20, 25–27, 29, 30, 34–36, 39
No	3, 5, 6, 9–11, 14, 17, 18, 21–24, 28, 31–33, 37, 38, 40

5.5 Appendix, Transient Stability Analysis in DIgSILENT

This section requires a basic understanding of DIgSILENT PowerFactory software. You must download the file (Chapter5.pfd) from the book's end-of-book attachments in order to follow this section.

Step 1:
    First, you need to check box 1 as shown in Figure 5.6, and then select **Dynamic Simulation** (No. 2).
Step 2: Determining the variables to be plotted
    *1. Right-click on the G1 generator.
    *2. See Figure 5.7. From the Define option, check the **Results for** RMS/EMT Simulation option.
    *3. In the table in Figure 5.8, double-click on the desired element icon.

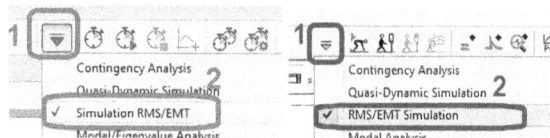

FIGURE 5.6
Viewing the **RMS/EMT Simulation** in v15.1 and v2021.

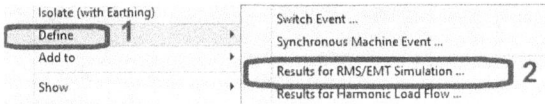

FIGURE 5.7
Defining generator variables for drawing dynamic curves.

FIGURE 5.8
Variable selection window.

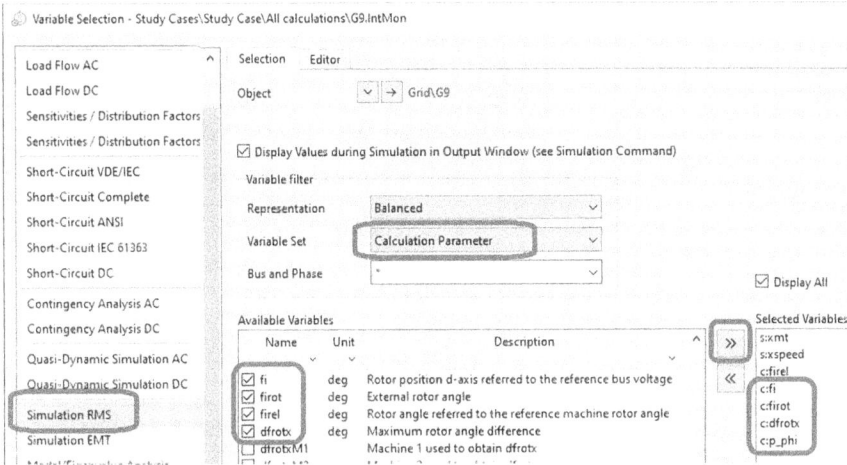

FIGURE 5.9
Display of variables of each generator to draw a dynamic curve.

*4. See Figure 5.9. "firel" is the parameter in question, and it can be found on the **Simulation RMT** page. By default, all you have to do is click the "OK" button.

*5. Click OK and return to the main network. Repeat steps (*1) to (*4) for the other two generators.

Step 3: Specify the type and time of fault

Right-click on Line36 and select **Short-Circuit Event** from Define. Set the time to 0.1 s and click OK. (Figure 5.10).

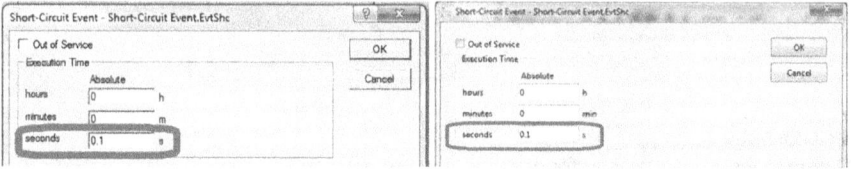

FIGURE 5.10
Specifying the short-circuit fault time.

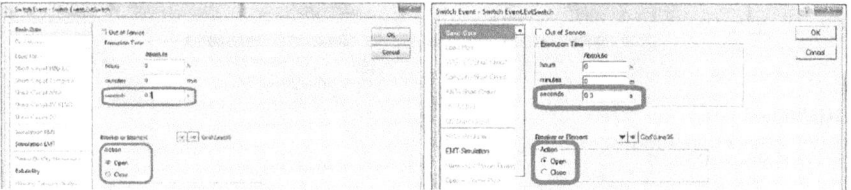

FIGURE 5.11
Specifying the switch event time.

Step 4:
Right-click on Line36 again and select **Switch Event** from the **Define** menu. In the Basic Data section, set the time to 0.3 s and set the **Action to Open**.

Step 5:
*Note: The icons ( ) provide a summary of the actions taken in steps 3 and 4. You must select the icon ( ) to modify the simulation elements and the icon ( ) to modify the short-circuit time. Press the reset button after adjusting any DIgSILENT parameters.

Step 6: Drawing a graphic display page

*1. In version 2016 and later, right-click on the+sign next to the edge of the main grid window (at the top of the figure) called **Grid**. (Grid × +)

*2. In the window that appears, select **Plot Page**. Choose a name and press **Execute**. (Creates a graphic page)

*3. At the top of the opened window, select the **Insert Plot** icon ( ), select the **Simulation RMS/ENT** (Curve plot) option, and click OK.

*4. Right-click on the figure and select **Edit**.

Step 7:
See Figure 5.12. Pay attention to the numbers.

*1. At the bottom of Figure 5.12, in the empty **Element** section (No. 1), double-click and select the generator G1 from the table that appears (No. 2).

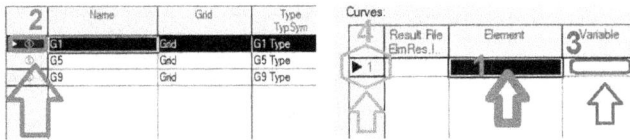

FIGURE 5.12
Adding variables to the simulation page.

*2. If (*1) does not work, you should run **Load Flow** and press the **Calculate Initial Condition** button.

*3. In the empty **Variables** section (No. 3), double-click and select one of the variables that appear. (Same **fire1**)

*4. Right-click on the edge of the first row of the table (No. 4) and select the **Insert Rows** option. (Do this twice)

*5. In the newly created rows, perform the clauses (*1) and (*3) for the generators G5 and G9 and press the **OK** option at the end.

Step 8:

*1. Now select the **Calculate Initial Condition** (🖈) or (🕒) option and press the Execute key. If you have followed the steps correctly; the network will not get an error.

*2. Now you can press the last key, i.e. Start Simulation (🔥) or (🕒) and specify the simulation stop time, for example 10 seconds, and press the **Execute** key.

*3. If the curves are not visible after the simulation is complete, use the Scale (⊷ I⁚) or (⊷ I) icons (top edge of the graphics screen).

Step 9: Changing the short-circuit clearing time to determine the critical clearing time

For any change, you must first press the reset button (🞂) or (🞃). Then, by pressing the simulation edit button (🞂) or (🞃), you can change the fault clearing time.

5.6 Summary

This chapter focuses on using DIgSILENT PowerFactory software for transient stability analysis, emphasizing the importance of Nonlinear Dynamics Methods in understanding power system behavior during disturbances. It specifically examines transient stability, which assesses a system's ability to maintain synchronism and voltage stability after disruptions. Engineers can simulate dynamic behavior to evaluate recovery capabilities, with the software allowing for the manipulation of over 2000 parameters to generate dynamic curves.

References

1. Eidiani, M., Buygi, M.O., Ahmadi, S., "CTV, complex transient and voltage stability: A new method for computing dynamic ATC," *International Journal of Power and Energy Systems*, 2006, 26(3), pp. 296–304, https://doi.org/10.2316/Journal.203.2006.3.203-3597.
2. Eidiani, M., Badokhty, M.E., Ghamat, M., Zeynal, H., "Improving transient stability using combined generator tripping and braking resistor approach," *International Review on Modelling and Simulations*, 2011, 4(4), pp. 1690–1699.
3. Eidiani, M., Baydokhty, M.E., Ghamat, M., Zeynal, H., Mortazavi, H. "Transient stability enhancement via hybrid technical approach," *2011 IEEE Student Conference on Research and Development*, 2011, pp. 375–380, https://doi.org/10.1109/SCOReD.2011.6148768.
4. Eidiani, M., "Atc evaluation by CTSA and POMP, two new methods for direct analysis of transient stability," *IEEE/PES Transmission and Distribution Conference and Exhibition*, 2002, vol. 3, pp. 1524–1529, https://doi.org/10.1109/TDC.2002.1176824.
5. Eidiani, M., "A reliable and efficient holomorphic approach to evaluate dynamic available transfer capability," *International Transactions on Electrical Energy Systems*, 2021, 31(11), p. e13031, https://doi.org/10.1002/2050-7038.13031.
6. Eidiani, M., "An efficient differential equation load flow method to assess dynamic available transfer capability with wind farms," *IET Renewable Power Generation*, 15, pp. 3843–3855, https://doi.org/10.1049/rpg2.12299
7. Eidiani, M., Buygi, M.O., Ahmadi, S., "CTV, complex transient and voltage stability: A new method for computing dynamic ATC," *International Journal of Power and Energy Systems*, 2006, 26(3), pp. 296–304, https://doi.org/10.2316/Journal.203.2006.3.203-3597.
8. Eidiani, M., Shanechi, M.H.M., "FAD-ATC: A new method for computing dynamic ATC," *International Journal of Electrical Power and Energy Systems*, 2006, 28(2), pp. 109–118, https://doi.org/10.1016/j.ijepes.2005.11.004.
9. Eidiani, M., Zeynal, H., Zakaria, Z., "Development of online dynamic ATC calculation integrating state estimation," *2022 IEEE International Conference in Power Engineering Application (ICPEA)*, 2022, pp. 1–5, https://doi.org/10.1109/ICPEA53519.2022.9744694.
10. Eidiani, M., "Online dynamic ATC computation with large-scale wind farms," *Electrical Engineering*, 2024, 106, pp. 5677–5684. https://doi.org/10.1007/s00202-024-02325-8.
11. Eidiani, M., Zeynal, H., Zadeh, A.K., Nor, K.M., "Exact and efficient approach in static assessment of Available Transfer Capability (ATC)," *2010 IEEE*

International Conference on Power and Energy, 2010, pp. 189–194, https://doi.org/10.1109/PECON.2010.5697580.

12. Eidiani, M., Yazdanpanah, D., "Minimum distance, a quick and simple method of determining the static ATC," *Journal of Electrical Engineering*, 2011, 11(2), pp. 95–101.

13. Eidiani, M., Asadi, S.M., Faroji, S.A., Velayati, M.H., Yazdanpanah, D. "Minimum distance, a quick and simple method of determining the static ATC," *2008 IEEE 2nd International Power and Energy Conference*, 2008, pp. 490–493, https://doi.org/10.1109/PECON.2008.4762524.

14. Eidiani, M., Zeynal, H. "Determination of online DATC with uncertainty and state estimation," *2022 9th Iranian Conference on Renewable Energy & Distributed Generation (ICREDG)*, 2022, pp. 1–6, https://doi.org/10.1109/ICREDG54199.2022.9804581.

15. Eidiani, M., Zeynal, H., Shaaban, M. "A detailed study on prevailing ATC methods for optimal solution development," *2022 IEEE International Conference on Power and Energy (PECon)*, 2022, pp. 299–303, https://doi.org/10.1109/PECon54459.2022.9988775.

16. Eidiani, M., "A new hybrid method to assess available transfer capability in AC–DC networks using the wind power plant interconnection," *IEEE Systems Journal*, 17(1), pp. 1375–1382, https://doi.org/10.1109/JSYST.2022.3181099.

17. Eidiani, M., Zeynal, H. "A fast holomorphic method to evaluate available transmission capacity with large scale wind turbines," *2022 9th Iranian Conference on Renewable Energy & Distributed Generation (ICREDG)*, 2022, pp. 1–5, https://doi.org/10.1109/ICREDG54199.2022.9804527.

18. Eidiani, M., Zeynal, H., Zakaria, Z. "An efficient holomorphic based available transfer capability solution in presence of large scale wind farms," *2022 IEEE International Conference in Power Engineering Application (ICPEA)*, 2022, pp. 1–5, https://doi.org/10.1109/ICPEA53519.2022.9744711.

19. Eidiani, M., Zeynal, H., Zakaria, Z. "An efficient method for available transfer capability calculation considering cyber-attacks in power systems," *2023 IEEE 3rd International Conference in Power Engineering Applications: Shaping Sustainability Through Power Engineering Innovation, ICPEA 2023*, 2023, pp. 127–130, https://doi.org/10.1109/ICPEA56918.2023.10093168.

20. Eidiani, M., Zeynal, H., "An effective method to determine the available transmission capacity with variable frequency transformer," *International Transactions on Electrical Energy Systems*, 2023, 2023, p. 8404284, https://doi.org/10.1155/2023/8404284.

21. Miranian, A., Rouzbehi, K., "Nonlinear Power System Load Identification Using Local Model Networks," *IEEE transactions on power system*, pp. 2872–2881, 2013.

22. Liu, Y., Raza, A., Rouzbehi, K., Li, B., Xu, D., Williams, B. W., "Dynamic resonance analysis and oscillation damping of multiterminal DC grids," *IEEE Access*, vol. 5, pp. 16974–16984, 2017.

6

Control Strategies for Power System Stability

6.1 Introduction

In this chapter, you will find a swift and concise explanation of how a Power System Stabilizer (PSS) functions in an electric power system. Also, the main ideas behind the PSS design, analysis, and application are covered in this chapter. Initially, a thorough analysis of the PSS design for a loaded generator connected to an infinite bus will be conducted. Firstly, without relying on a PC, the reader can simply follow every step of PSS design. After that, all PSS design formulas are consolidated into a single MATLAB® program for the sake of convenience. Finally, DIgSILENT PowerFactory simulations demonstrate PSS performance in a 9-bus power grid and a large real grid. How to use the software and the possibility for further work are explained in the appendix. For more detailed work and further work, you can refer to the references [1–15] at the end of this chapter.

6.2 Nomenclature and Symbols

f_0	Grid frequency
R, X	Line impedance
G, B	Load admittance
K_A, T_A	Exciter gain and time-constants
K_G, T_G	Governor gain and time-constants
K_T, T_T	Turbine gain and time-constants
K_3, T_3, T'_{do}	Field circuit gain and time-constants ($T_3 = K_3 T'_{do}$)
M, D	Generator damping and inertia
x_d, x'_d, x_q, r	Generator parameters
K_1-K_6	Heffron-Phillips model gains
V_t	Terminal voltage
V_∞	Infinite bus voltage
ω_n	Oscillation frequency

DOI: 10.1201/9781003590514-6

T_1	Phase delay between V_{ref} and E'
T_2	Time constant in the phase compensator
T	Washout filter or reset time-constants
$D = D_E + D_m$	Electric and mechanical damper
ζ_n	Damping coefficient $(D_E/(2M\omega_n))$
K_C	The compensation gain in PSS

Initial condition: $V_q^0, V_d^0, I_q^0, I_d^0, \delta^0, E_0'$

Intermediate parameters to calculate PSS parameters:
$C_1, C_2, R_1, R_2, X_1, X_2, X_3, X_4, Z_e, Y_q, Y_d, F_d, F_q, A_{dq}$

The main formula of salient pole synchronous machine:

$$E' = V_t + rI_a + jx_d I_d e^{j\delta} + jx_q I_q e^{j\delta}$$

6.3 Introduction to PSS

The reliable and efficient operation of a power system hinges on its ability to maintain stability. This stability refers to the system's capability to return to a steady-state operating condition following a disturbance. PSS play a crucial role in safeguarding this stability, particularly by mitigating low-frequency oscillations in the rotors of generators.

Power systems inherently operate with a narrow stability margin. Even minor disturbances can trigger instability at specific operating points, leading to cascading outages and widespread power disruptions. These disturbances can originate from various sources, including sudden changes in load demand, faults on transmission lines, and loss of generation units. Instability manifests as sustained oscillations in the rotor speeds of generators. These oscillations, typically in the range of 0.1–2 Hz, deviate the generators from their synchronous speed and can lead to voltage instability and tripping of circuit breakers.

PSS acts as a safeguard against these detrimental oscillations. Their primary function is to **dampen** these oscillations, ensuring the rotors return to their synchronous speed quickly and smoothly after a disturbance. Damping essentially refers to the gradual reduction in the amplitude of the oscillations over time.

PSS achieves this damping effect by injecting an **auxiliary stabilizing signal** into the **Automatic Voltage Regulator (AVR)** of the generator. The AVR is a vital control element within the generator's excitation system, responsible for regulating the generator's terminal voltage.

By strategically modifying the generator's excitation based on the injected signal, PSS influences the generator's real power output and, consequently, its rotor speed. This targeted control action helps dampen the oscillations and restore stable operation.

Traditionally, the **grid frequency** or **rotor speed** of the generator serves as the primary input signal for the PSS. This information allows the PSS to detect and respond to any deviations from the desired synchronous speed.

However, in modern power systems with multiple PSS-equipped generators, using the power signal as an alternative input can be advantageous. This approach **prevents interaction** between individual PSS units. The rationale behind connecting the PSS output signal to the exciter lies in the fundamental differences between the exciter and the governor. The **governor** is a slower-acting device primarily concerned with regulating the generator's real power output (megawatts). It typically involves mechanical connections for control. The **exciter** operates at a much faster speed and regulates the generator's field voltage (kilowatts). Electrical connections facilitate its control. Therefore, due to the exciter's faster response and higher operational speed, it provides a more effective point for injecting the PSS output signal for influencing the generator's real power output and damping rotor oscillations. Figure 6.1a shows the signal connection between PSS and AVR and the exciter generator.

Here it is necessary to give you a clear idea of PSS itself. PSS is a circuit and control module that is made of electronic components. These electronic parts manipulate the control signals and increase, delay, and accelerate the signal, causing the stability of the network. A simple figure of the control block diagram of a PSS is shown in Figure 6.1b, which will be discussed in detail in the next sections.

To continue, you must review the materials related to linear dynamics, especially the Heffron-Phillips model in previous chapters. The PSS block integrated with the Heffron-Phillips model can be shown in Figure 6.1c. The reader should be aware of how to linearize a generator connected to an infinite bus and calculations from (k_1) to (k_6) so that a PSS can be designed for this network.

Two inputs (ΔV_{ref} and ΔP_{ref}) are derived from the reactive power and optimal power flow calculation sent by the transmission system operator (TSO), respectively (See Figure 6.1c). Low-frequency oscillations (associated with the slow variations in the electrical quantities within the power system ranging from 0.1 to 3 Hz) are caused by connecting two distinct networks—normally, one serving as a producer and the other as a consumer, even with some small distributed generations—through a single line. Power oscillation can occur within a network with other parts or among multiple power plants. In fact, all network variables oscillate following the power oscillation.

In the following, it discusses significant issues pertaining to PSS's design and operation.

6.3.1 PSS Design

In this section, we review the PSS design algorithm without paying attention to the details of the PSS design steps. In short, this design tries to

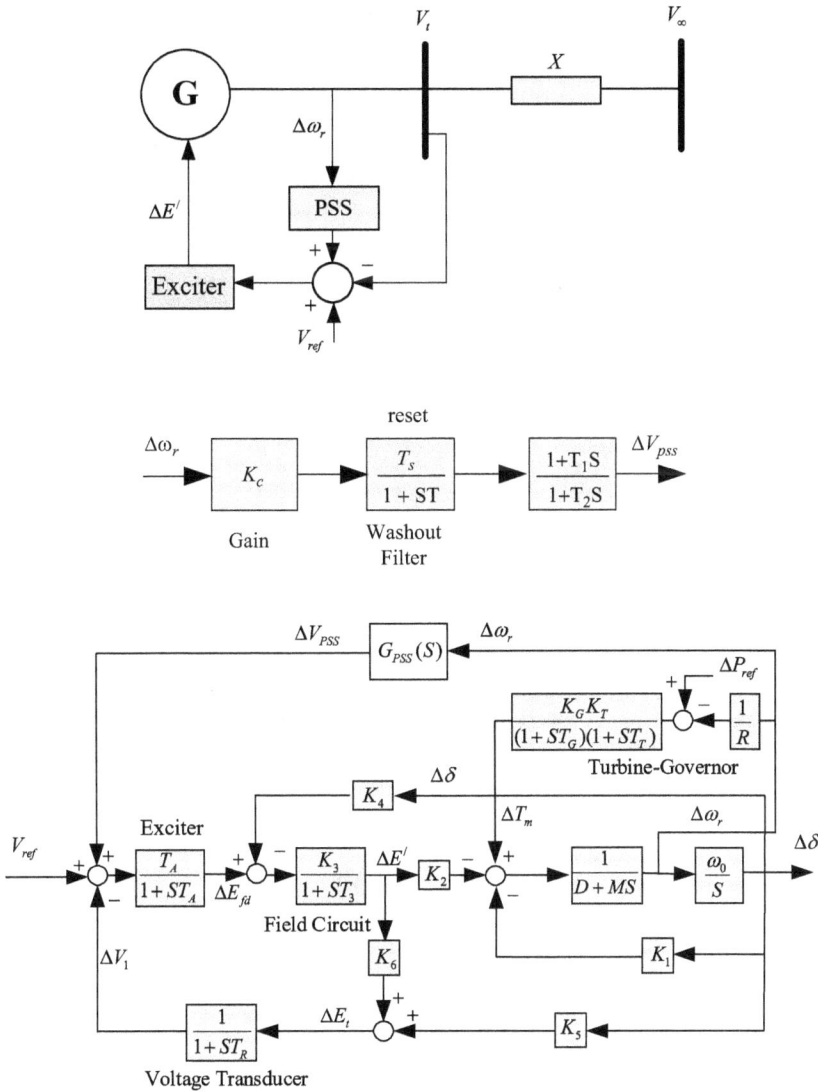

FIGURE 6.1

(a) Generator with AVR and PSS. (b) Block diagram representation of PSS. (c) PSS integrated into the Heffron-Phillips model.

generate a signal to remove the unstable poles by adding a zero to the network. The PSS design algorithm is briefly defined as follows:

- Determining the frequency of oscillation of the mechanical loop, or eigenvalues of the system.

- Determining the phase delay between V_{ref} and E'.
- Designing a phase compensator with a high delay.
- Determining the compensation gain.
- Determining of reset block for non-operation of PSS in normal oscillations.

6.3.2 PSSs Interference

Although it seems that the network should become stable with the addition of PSS if there are many PSS generators, these PSSs will interfere with each other, so you need to think of a solution to solve this problem. In fact, the PSS of each PSS-equipped generator may amplify the oscillation of other generators if they are oscillating at different frequencies. An approach can be designing the entire network at once, a practice that is often impractical, mainly because of two reasons:

- Orders for generators are placed based on various technologies and at different times.
- Generators are made by various vendors.

In real life, PSS interference can be eliminated using power signals rather than speed. Further exploration of these methods is encouraged. A list of relevant research papers and references can be found at the end of this book [1–9].

6.3.3 Network Reduction Methods

Network reduction, to put it briefly, is the process of placing a set of equations in a manner such that their collective impact on the system imitates that of the main network, rather than dealing with multiple large networks. This reduced network is usually much simpler to study. Power transmission capacity estimation, PSS design and analysis, and transient stability analysis all extensively employ power network reduction. The following succinctly lists the network reduction methods.

Finally, it should be said that PSS design and network reduction methods are closely related, which is why these methods are reviewed here. To learn more about these methods, readers can explore the relevant research papers and references listed at the end of this book [1–9]. The following succinctly lists network reduction methods.

- The **inertia technique**.
- The **area division technique** (partitioning an area into three; major, typical, and low-impact sections).
- **Model change technique** (impedance, linear, and non-linear models).

- **The modal approach**, in which a pair of eigenvalues is referred to as a mode. This technique analyzes up to six modes and removes high-frequency modes.
- The **coherency method** models machines that oscillate in unison using a single machine. Coherency has a high computation cost and is a challenging method.
- The network is estimated using the **estimation method** based on transmission line currents and voltages. You can use this method online.

6.3.4 Subsynchronous Resonance

Each component of the low-pressure and high-pressure turbines, exciter, and generator has elasticity and oscillates in relation to one another. When there is no resonance, low-frequency oscillations are suppressed and eliminated. However, in the case of resonance, the low-frequency oscillations drastically increase over time. In the subsynchronous resonance (SSR) mode, the rotor oscillates at the subsynchronous frequency. Then a current with the same frequency is generated in the stator. Due to the presence of a suitable capacitor in the network, this current is intensified in the network. This current enters the generator again at the subsynchronous frequency. The existence of a total capacitor in the network equal to the network reactors at this frequency is the primary cause of subsynchronous resonance.

For a deeper dive into these methods, consult the research papers and references provided at the book's conclusion [1–9].

6.3.5 Determining Frequency Modes and Their Elimination Methods

As was said before, the identification of generator oscillation modes is the most important parameter in PSS design, which is summarized here. Applying a small shock (for instance a short circuit) to the generator while operating at low speed is the most effective, affordable, and secure method of identifying generator oscillation modes. More naturally, a slight phase difference during the grid synchronization can also be used to check the generator. It is also possible to identify the generator's frequency modes by switching the network capacitors. The most precise approach involves feeding the generator an AC excitation current with oscillating frequencies and monitoring the generator's output.

In summary, the following techniques can be used to remove oscillating modes: Frequency filtering, rotor damping, short circuit capacitor during the resonance, and exciter-based generator control.

Readers interested in delving deeper into these methods can consult the relevant research papers and references included in the bibliography [1–9].

6.4 A Simple PSS Design

In this section, briefly and attractively, a PSS is designed without the need for a computer for a loaded generator connected to an infinite bus. This section and "Section 6.5" are very suitable and attractive for university students and researchers, but if the reader of the book is an industrial engineer, the reader can ignore these two sections and go directly to "Section 6.6".

(See Figure 6.2) The following information is known in this network: Grid frequency f_0, Generator $(M, T'_{do}, x_d, x'_d, x_q, r)$, Line (R, X), Load (G, B), Exciter (K_A, T_A), Initial condition (V_t^0, P_e^0, Q_e^0), also, suppose these values are known (ζ_n, T_2, T). The goal here is to determine the parameters (K_C, T_1) in the PSS block as follows: (See Figure 6.1b)

$$G_{PSS}(s) = \left(\frac{sT}{1+sT}\right)\left(\frac{K_C(1+sT_1)}{1+sT_2}\right) \tag{6.1}$$

In fact, all the contents of this section and "Section 6.5" are written to determine these two parameters (K_C, T_1).

We need to determine the following parameters in the correct order using the known information. First, Appendix 1 (Section 6.9) provides proof of identifying the network's initial conditions. By knowing the previously mentioned parameters (network information) and determining the initial conditions Appendix 1 (Section 6.9), any researcher can prove the following relationships and follow the PSS design calculations.

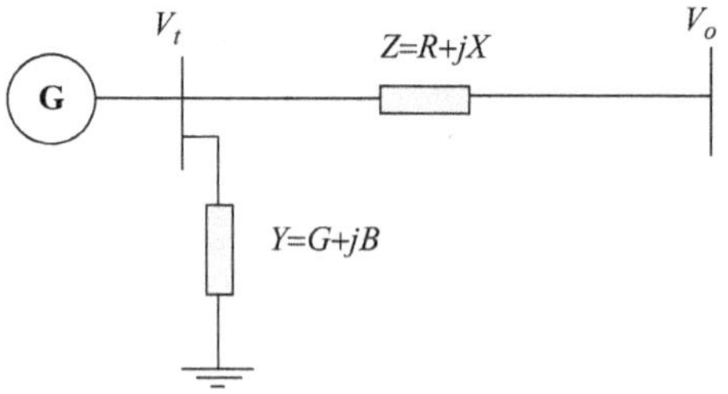

FIGURE 6.2

A loaded generator connected to an infinite bus.

6.4.1 Simplification of Line and Load Parameters

The parameters and relationships listed below are completed in numerical order.

1. $C_1 = 1 + RG - XB$
2. $C_2 = RB + GX$
3. $R_1 = R + C_1 r - C_2 x'_d$
4. $R_2 = R + C_1 r - C_2 x_q$
5. $X_1 = X + C_2 r + C_1 x_q$
6. $X_2 = X + C_2 r + C_1 x'_d$
7. $Z_e^2 = R_1 R_2 + X_1 X_2$
8. $Y_q = \dfrac{R_1 C_1 + C_2 X_2}{Z_e^2}$
9. $Y_d = \dfrac{R_2 C_2 - C_1 X_1}{Z_e^2}$

6.4.2 Determining Initial Conditions for the System

10. $A_{dq} = \left(\dfrac{-x_q P_e^0 + r Q_e^0}{\left(V_t^0\right)^2 + x_q Q_e^0 + r P_e^0} \right)$

11. $V_q^0 = \dfrac{V_t^0}{\sqrt{1 + A_{dq}^2}}$

12. $V_d^0 = A_{dq} V_q^0$

13. $I_q^0 = \left(\dfrac{P_e^0 V_q^0 + Q_e^0 V_d^0}{\left(V_t^0\right)^2} \right)$

14. $I_d^0 = \left(\dfrac{P_e^0 V_d^0 - Q_e^0 V_q^0}{\left(V_t^0\right)^2} \right)$

15. $E_0' = V_q^0 + r I_q^0 - x_d' I_d^0$

16. $X_3 = R I_q^0 - X I_d^0 - C_1 V_q^0 + C_2 V_d^0$

17. $X_4 = R I_d^0 + X I_q^0 - C_1 V_d^0 - C_2 V_q^0$

18. $\delta^0 = \tan^{-1}\left(\dfrac{X_4}{X_3} \right)$

19. $V_\infty = \dfrac{X_4}{\mathrm{Sin}(\delta_0)}$

20. $F_q = \dfrac{V_\infty(R_1\sin(\delta^0) + X_2\cos(\delta^0))}{Z_e^2}$

21. $F_d = \dfrac{V_\infty(R_2\cos(\delta^0) - X_1\mathrm{Sin}(\delta^0))}{Z_e^2}$

6.4.3 Determining K_1 to K_6

22. $K_1 = F_q(E_0') + (x_d' - x_q)(F_d I_q^0 + I_d^0 F_q)$

23. $K_2 = I_q^0 + Y_q(E_0') + (x_d' - x_q)(Y_q I_d^0 + Y_d I_q^0)$

24. $K_3 = \dfrac{1}{1 - (x_d - x_d')Y_d}$

25. $K_4 = (x_d' - x_d)F_d$

26. $K_5 = F_d\, x_d'\, \dfrac{V_q^0}{V_t^0} - F_q\, x_q\, \dfrac{V_d^0}{V_t^0}$

27. $K_6 = \dfrac{V_q^0}{V_t^0} + Y_d\, x_d'\, \dfrac{V_q^0}{V_t^0} - \dfrac{Y_q\, x_q\, V_d^0}{V_t^0}$

6.4.4 Determining the Parameters of the PSS Control Block

28. $\omega_n = \sqrt{\dfrac{2\pi(f_0)K_1}{M}}$

29. $|G_E| = \left| \dfrac{K_A K_3}{(1 + j\omega_n T_A)(1 + j\omega_n T_{do}' K_3) + K_A K_3 K_6} \right|$

30. $\angle G_E = \angle\left(\dfrac{K_A K_3}{(1 + j\omega_n T_A)(1 + j\omega_n T_{do}' K_3) + K_A K_3 K_6} \right)$

31. $T_1 = \dfrac{\tan\left(\tan^{-1}(\omega_n T_2) - \angle G_E\right)}{\omega_n}$

32. $|G_C| = \left| \dfrac{1 + j\omega_n T_1}{1 + j\omega_n T_2} \right|$

33. $K_C = \dfrac{2\zeta_n \omega_n M}{K_2 |G_E||G_C|}$

6.5 An Example of a Simple PSS

In this section, two numerical examples are presented based on the relations and formulas of the previous section that you accepted without proof. In almost all references, the generator resistance is ignored in the calculation of PSS parameters, but in example (a) the generator resistance is included and the formulas of the previous section are written in the presence of resistance. To compare relations with other references [1–9], example (b) is the famous example of without resistance.

Now you can design a PSS for Figure 6.2 with the following data.

a. $f = 60$ Hz, $T_2 = 0.15$ s, $T = 2$ s, $\xi_n = 0.25$ pu, $M = 10$ s, $T'_{do} = 9.53$ s

$x_d = 0.976$ pu, $x'_d = 0.253$ pu, $x_q = 0.576$ pu, $r = 0.005$ pu

$R = 0.050$ pu, $X = 0.944$ pu, $G = 0.276$ pu, $B = 0.267$ pu

$K_A = 50$ pu, $T_A = 0.05$ s, $P_{e0} = 1.05$ pu, $Q_{e0} = 0.019$ pu, $V_{t0} = 1.07$ pu

b. $f = 60$ Hz, $T_2 = 0.1$ s, $T = 3$ s, $\xi_n = 0.3$ pu, $M = 9.26$ s, $T'_{do} = 7.76$ s

$x_d = 0.973$ pu, $x'_d = 0.190$ pu, $x_q = 0.55$ pu, $r = 0$ pu

$R = -0.034$ pu, $X = 0.997$ pu, $G = 0.249$ pu, $B = 0.262$ pu

$K_A = 50$ pu, $T_A = 0.05$ s, $P_{e0} = 1$ pu, $Q_{e0} = 0.015$ pu, $V_{t0} = 1.05$ pu

Researchers can do all the relations without the need for a computer. But here, for simplicity, the relations are given in a MATLAB language file according to Table 6.1, so that the order of calculations can be seen. First, system (a) is analyzed.

Table 6.2 shows the outcomes derived from the completed calculations: Example output (- example (a)):

Therefore, the PSS diagram block for example (a) is completed as follows: (Figure 6.1b and equation (6.1)):

$$G_{PSS}(s) = \left(\frac{2s}{1+2s}\right)\left(\frac{0.4129(1+12.1645\,s)}{1+0.15\,s}\right) \qquad (6.2)$$

Now, as in example (a), the output of system (b) is specified in Table 6.3.

Therefore, the PSS diagram block for example (b) is completed as follows (Figure 6.1b and equation (6.1)):

$$G_{PSS}(s) = \left(\frac{3s}{1+3s}\right)\left(\frac{7.0908(1+0.685\,s)}{1+0.1\,s}\right) \qquad (6.3)$$

TABLE 6.1

Program Written in MATLAB (Section 6.5-example (a))

```
clc;clear;              va0=1.07;                    XX2=X*iq0+R*id0-c1*vd0-c2*vq0
% Input Data           c1=1+R*G-X*B                 delta0=-atan(XX2/XX1)
f=60;                  c2=R*B+G*X                   v0=XX2/sin(delta0)
t2=0.15;               R1=R+c1*r-c2*xpd             fq=v0/Ze2*(R1*sin(delta0)+X2*cos
eta=0.25;              R2=R+c1*r-c2*xq                 (delta0))
M=10;                  X1=X+c2*r+c1*xq              fd=v0/Ze2*(-X1*sin(delta0)+R2*cos(d
tdo=9.53;              X2=X+c2*r+c1*xpd                elta0))
xd=0.976;              Ze2=R1*R2+X1*X2              k1=fq*ep0+(xpd-xq)*(fd*iq0+id0*fq)
xpd=0.253;             yq=(R1*c1+c2*X2)/Ze2         k2=iq0+yq*ep0+(xpd-xq)*(yq*id0+yd
xq=0.576;              yd=(-c1*X1+c2*R2)/Ze2           *iq0)
ka=50;                 Adq=(-xq*p0+r*q0)/           k3=1/(1-(xd-xpd)*yd)
ta=0.05;                 (va0^2+xq*q0+r*p0)         k4=(xpd-xd)*fd
R=0.050;               vq0=va0/(1+Adq^2)^0.5        k5=fd*xpd*vq0/va0-fq*xq*vd0/va0
X=0.944;               vd0=Adq*vq0                  k6=vq0/va0+yd*xpd*vq0/
G=0.276;               iq0=(p0*vq0+q0*vd0)/va0^2       va0-yq*xq*vd0/va0
B=0.267;               id0=(p0*vd0-q0*vq0)/va0^2    wn=sqrt(k1*2*pi*f/M)
r=0.005;               ep0=vq0+r*iq0-xpd*id0        GE=ka*k3/((1+j*wn*ta)*(1+j*wn*tdo*
p0=1.05;               XX1=R*iq0-X*id0-c1*vq0+c2*vd0   k3)+ka*k3*k6)
q0=0.019;                                           abGE=abs(GE)
                                                    anGE=angle(GE)
                                                    t1=tan(atan(wn*t2)-anGE)/wn
                                                    abGC=abs((1+j*wn*t1)/(1+j*wn*t2))
                                                    Kc=2*eta*wn*M/(k2*abGE*abGC)
```

TABLE 6.2

Output of the Table 6.1

c1 = 0.7618	iq0 = 0.8621	k3 = 0.6682
c2 = 0.2739	id0 = −0.4690	k4 = 0.5933
R1 = −0.0155	ep0 = 1.0720	k5 = −0.1150
R2 = −0.1040	XX1 = −0.3724	k6 = 0.7834
X1 = 1.3841	XX2 = 0.9070	wn = 4.5389
X2 = 1.1381	delta0 = 1.1812	GE = 0.5408 − 0.7642i
Ze2 = 1.5769	v0 = 0.9805	abGE = 0.9362
yq = 0.1902	fq = 0.2599	anGE = −0.9549
yd = −0.6867	fd = −0.8207	t1 = 12.1645
Adq = −0.5208	k1 = 0.5465	abGC = 45.6474
vq0 = 0.9490	k2 = 1.2861	Kc = 0.4129
vd0 = −0.4942		

TABLE 6.3

Output of System (b)

c1=0.7303	iq0=0.8471	k3=0.6584
c2=0.2393	id0=−0.4354	k4=0.6981
R1=−0.0795	ep0=1.0237	k5=−0.0955
R2=−0.1656	X3=−0.3934	k6=0.8159
X1=1.3987	X4=0.9745	wn=4.7067
X2=1.1358	delta0=1.1871	GE =0.6751−0.7384i
Ze2=1.6017	v0=1.0509	abGE=1.0005
yq=0.1335	fq=0.2306	anGE=−0.8302
yd=−0.6625	fd=−0.8916	t1=0.6850
Adq=−0.4952	k1=0.5441	abGC=3.0544
vq0=0.9410	k2=1.2067	Kc=7.0908
vd0=−0.4659		

6.6 PSS in DIgSILENT PowerFactory

In this section, it is assumed that the reader is familiar with the funda-mentals of DIgSILENT PowerFactory. You can download the developed DIgSILENT source files from the appendix. Let us examine a 9-bus network (Figure 6.3). Assume that a three-phase short circuit fault has occurred at time (1 s) in the middle of transmission line 3–6. If the fault is cleared by opening lines 3–6 in the allotted time of 1.3300652 s, Figure 6.4 illustrates the generator's angle and shows the stability boundary.

If the fault is cleared by opening lines 3–6 in the allotted time of 1.3300653 s, Figure 6.5 illustrates the generator's angle and shows the tran-sient instability of the network. We now add a governor, an exciter, and a PSS with the following specifications to generator #5.

Governor, Exciter, and PSS information added in the previous section are given in Tables 6.4–6.7. Also, the block diagram of these controllers is given in Figures 6.6–6.8.

As demonstrated in Figure 6.5, the network is unstable if the fault-clear-ing time exceeds 1.3300653 s. The angle of generators with the same short circuit and generator # 5's PSS, governor, and exciter are shown in Figure 6.9. It is easy to see that PSS can easily stabilize transient unstable networks.

Now we can contemplate different scenarios. In the first scenario, if the PSS is moved to different generators, what effect does it have on the net-work transient stability? Also, if PSS is located on all generators, does it have a good effect on the stability of the network or not? In the second scenario, it shows the effect of changing PSS parameters on stability.

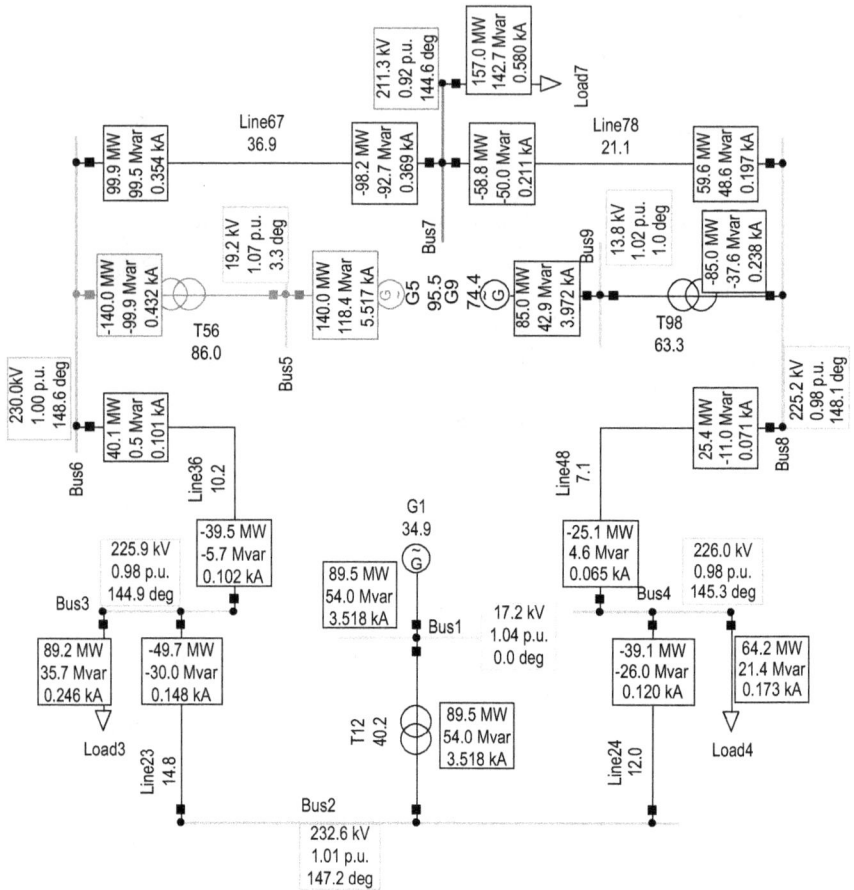

FIGURE 6.3
A simple 9-bus network.

6.6.1 Scenario 1: Relocation of PSS

In Figure 6.10, the PSS is on generator #9, and in Figure 6.11, the PSS is on generator #1. The PSS is operating on all three generators in Figure 6.12.

Figures 6.9–6.12 can now be compared. As an important result, the location of the PSS matters, and the stability rate and the amount of oscillation change. Also, by comparing Figure 6.12 with other outputs, it can be concluded that the improvement of network performance is not always associated with a large number of PSS. The importance of coordination between PSSs can also be seen here.

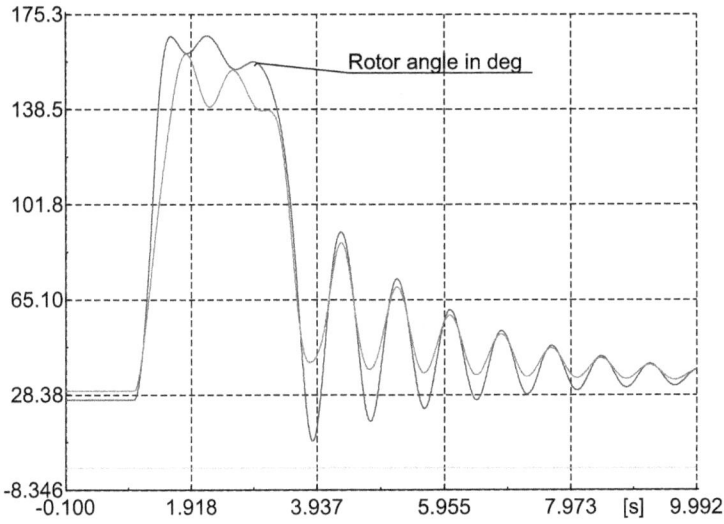

FIGURE 6.4
The angle of generators in the stability boundary.

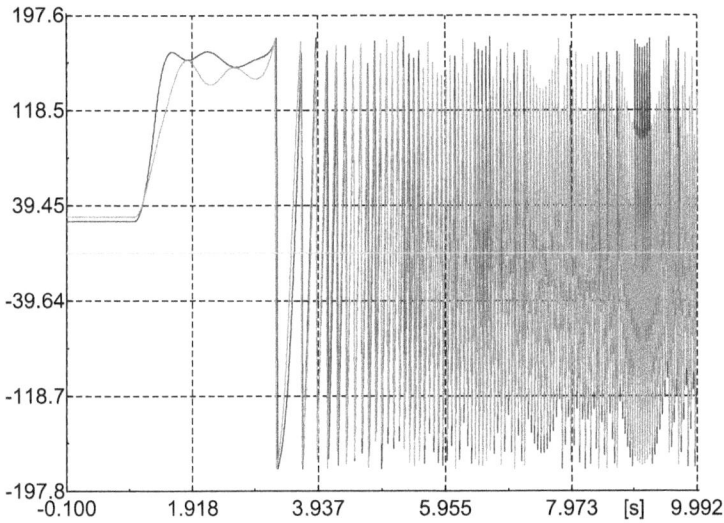

FIGURE 6.5
The angle of generators in the transient instability.

TABLE 6.4

Specifications of a Governor, an Exciter, and a PSS

Gen # 5	Governor (HYGOV)	Exciter (IEEET1)	PSS (Stab1)
Specifications	Table 6.5	Table 6.6	Table 6.7
Block diagram	Figure 6.6	Figure 6.7	Figure 6.8

TABLE 6.5

Specifications of a Governor (HYGOV)

R	0.04	r	0.5
T_r	8.408	T_f	0.05
T_g	0.5	T_w	0.496
A_t	1.15	P_{turb}	0
D_{turb}	0	Q_{nl}	0.08
G_{min}	0	Q_{nl}	0
V_{elm}	0.2	G_{max}	1

TABLE 6.6

Specifications of an Exciter (IEEET1)

T_r	0.028	K_a	175
T_a	0.03	K_e	1
T_e	0.266	K_f	0.0025
T_f	1.5	E_1	4.5
S_{e1}	1.5	E_2	6
V_{min}	−12	V_{max}	12

TABLE 6.7

Specifications of a PSS (Stab1)

K	100
T	10
T_2	0.1
T_4	0.05
T_1	0.1
T_3	0.05
H_{LIM}	0.05

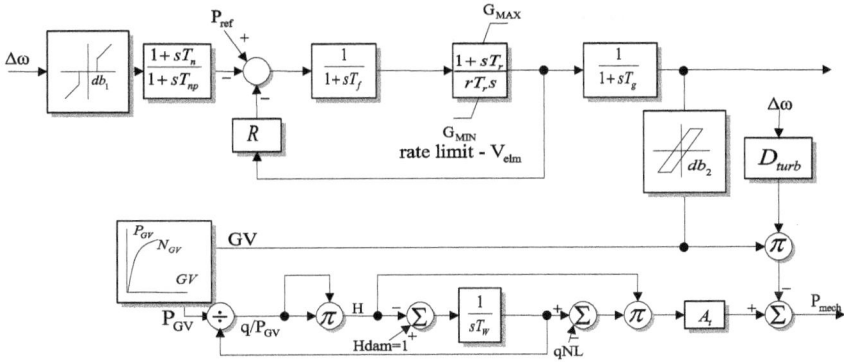

FIGURE 6.6
Governor block diagram presentation (HYGOV).

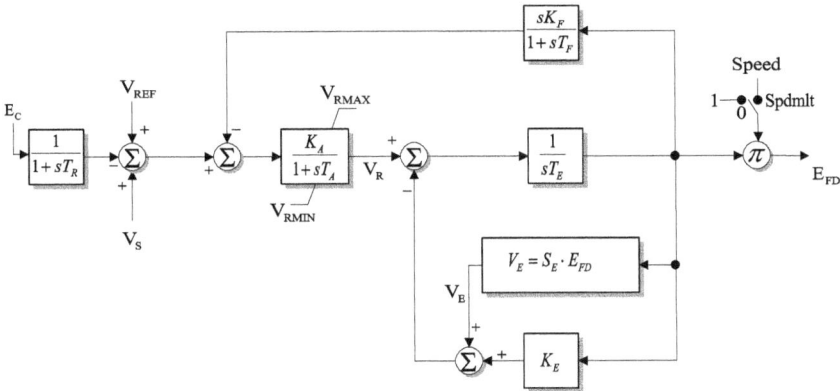

FIGURE 6.7
Exciter block diagram presentation (IEEET1).

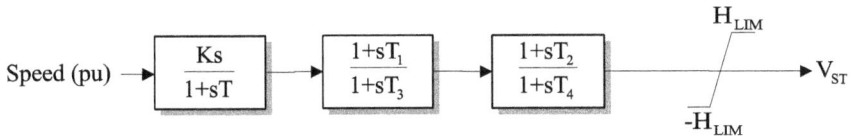

FIGURE 6.8
Block diagram of the PSS (Stab1).

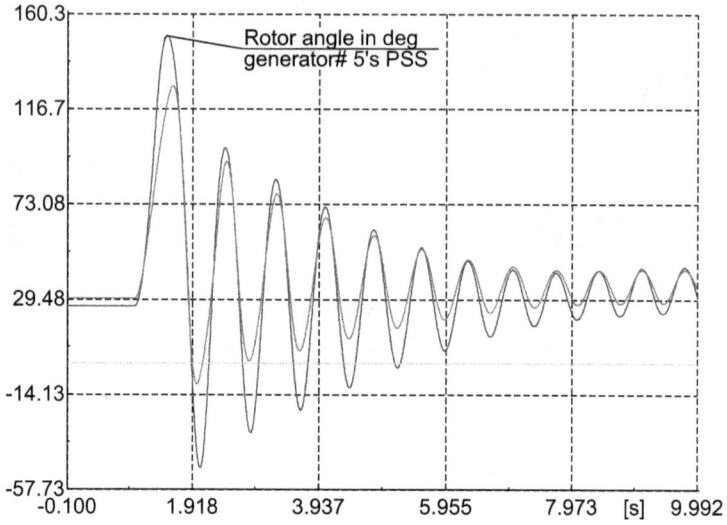

FIGURE 6.9
The angle of generators with generator # 5's PSS, governor, and exciter.

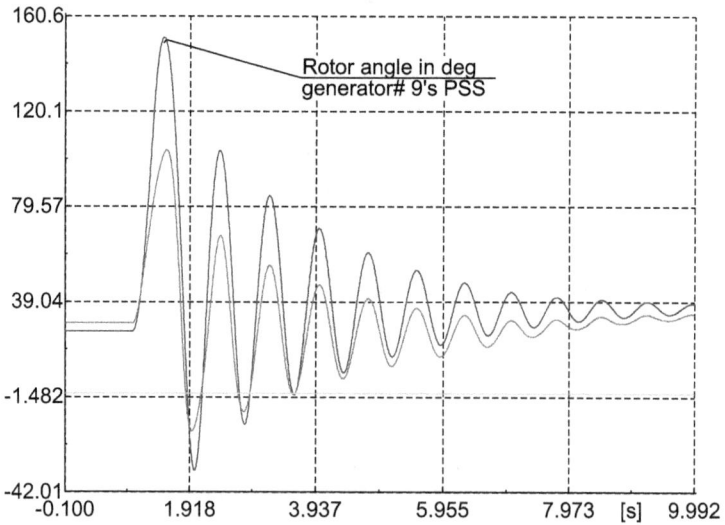

FIGURE 6.10
The angle of generators with generator# 9's PSS, governor, and exciter.

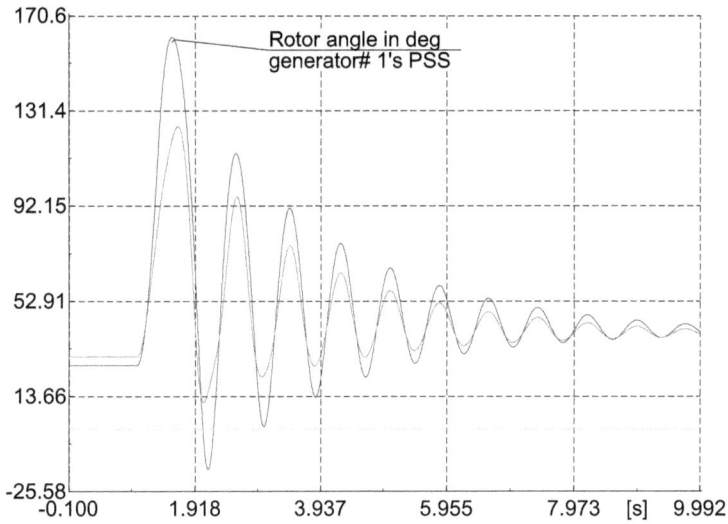

FIGURE 6.11
The angle of generators with generator #1's PSS, governor, and exciter.

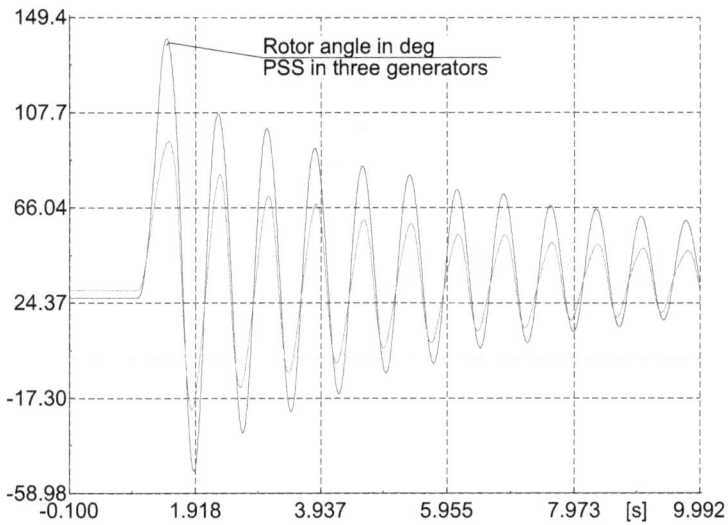

FIGURE 6.12
The angle of generators with the presence of PSS in three generators.

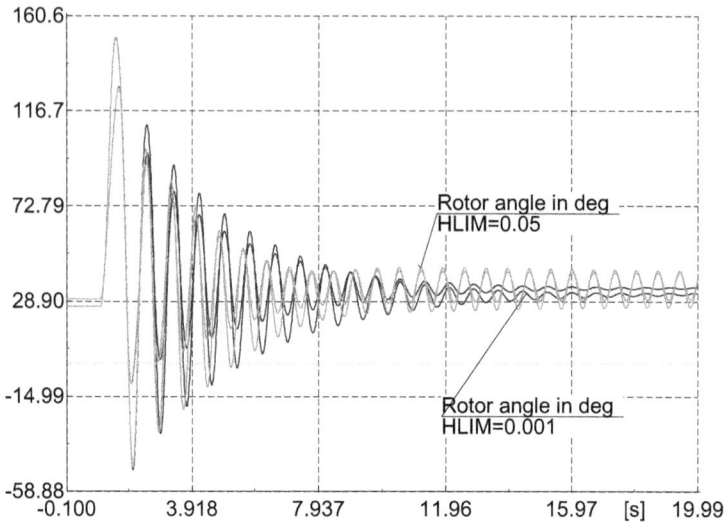

FIGURE 6.13
The angle of generators with changing PSS parameters.

6.6.2 Scenario 2: Changing PSS Parameters

In this scenario, the parameter (HLIM) is changed from HLIM=0.05 to HLIM=0.001 and the results can be seen in Figure 6.13. It is easy to see that reducing the HLIM parameter increases the initial angle fluctuations and decreases the final angle fluctuations of the generators, which should be considered in the design. Next, researchers can investigate the effects of changing other PSS parameters on transient stability.

6.7 PSS Analysis in Real Network

In this part, the analysis of the performance of PSS in a province is carried out from the Iranian network. This large network has 1900 buses and terminals, 38 three-winding transformers, 689 important loads, 1000 two-winding transformers, and 13,000 transmission lines. 66 generators in the province and a total of 350 generators have been modeled in this network. Figure 6.14 shows a part of the 400 kV network along with the generators of the province network. For those interested, the network data file can be downloaded in DIgSILENT PowerFactory format from the book's appendix. This allows you to replicate the analysis steps.

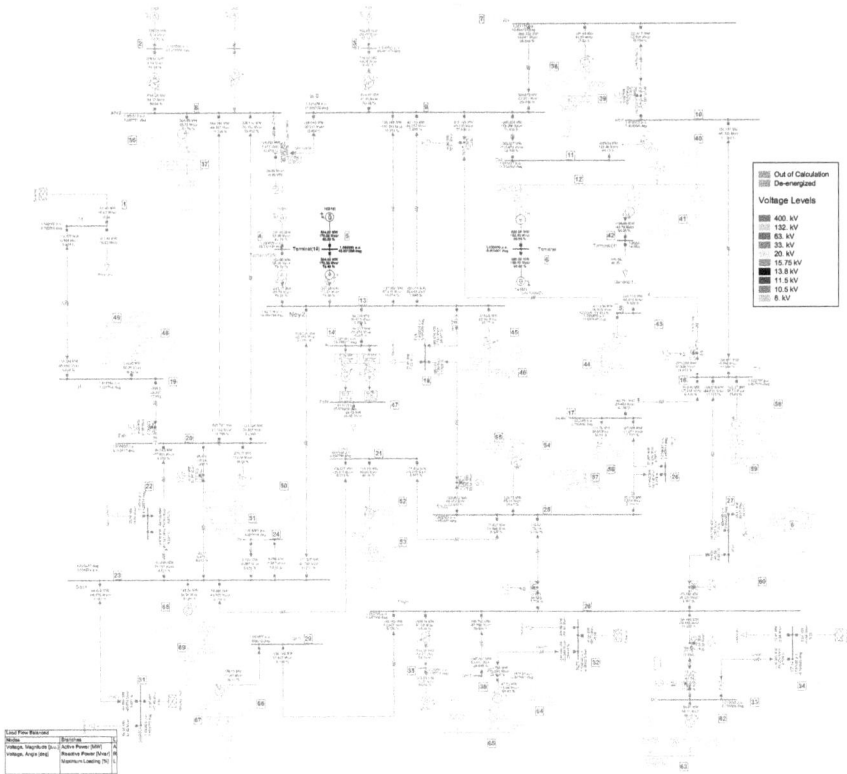

FIGURE 6.14
A part of the 400 kV network along with the generators of the Khorasan province network.

6.7.1 Network Without PSS

In this section, it is assumed that a short circuit occurs in the 400 kV transmission line between Bus 8 and Bus 20, 70 km long, and the short circuit is resolved by cutting the breakers on both sides of the transmission line. As the duration of the fault changes, the network will become stable and unstable in terms of transient stability. Figures 6.15 and 6.16 show two stable (clear time=0.225 second) and unstable (clear time=0.25 second) states of the network (for convenience, the angles of all generators are not shown in the figures). The critical clearing time at this work point is 225 ms.

In the next section, the effect of adding an arbitrary PSS on network stability is investigated.

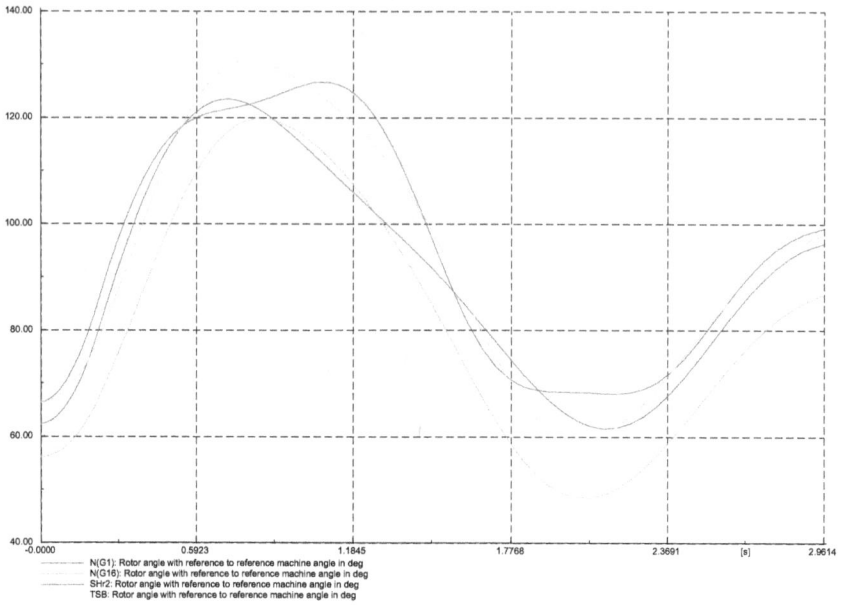

FIGURE 6.15
Angle generators in the transient stable state (clear time=0.225 second).

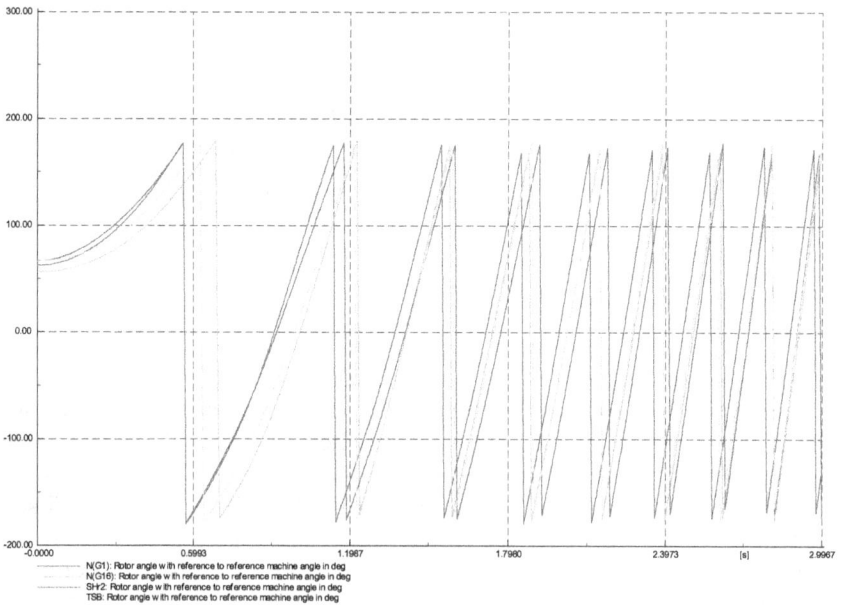

FIGURE 6.16
Angle generators in the transient unstable state (clear time=0.25 second).

6.7.2 Network with PSS

The effect of PSS is examined for the actual network in this section. According to Appendix 2 (Section 6.10), two PSSs, governors, and exciters have been added for two distinct generators, Bus 2 and Bus 4, in this network. It is anticipated that the presence of PSS will have a greater impact on Bus 2 because the short circuit is closer to it. Simulation in an unstable state has confirmed this prediction (see Figure 6.16).

The unstable network diagram when PSS is connected on bus 2 is shown in Figure 6.17. It is evident that the network remains stable in these circumstances. Figure 6.18 is the unstable network diagram when PSS is connected to bus 4. It can be seen that the network is not stable in this condition. Furthermore, the network has not been stabilized by this PSS, as expected.

In summary, you can see how crucial it is to be positioned as optimally as possible within the network. For additional research on the best displacement and PSS values to increase network stability, readers can consult the relevant references at the book's conclusion [1–9].

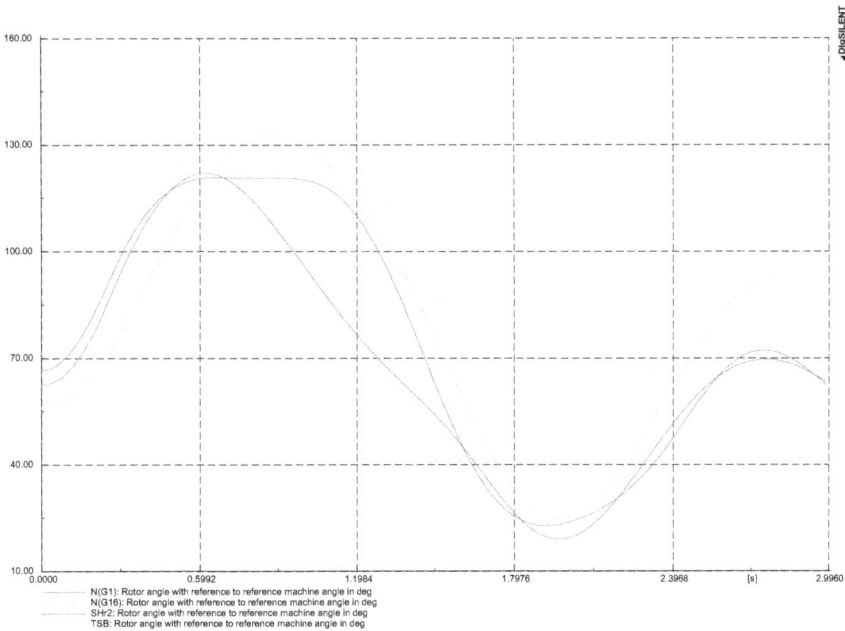

N(G1): Rotor angle with reference to reference machine angle in deg
N(G16): Rotor angle with reference to reference machine angle in deg
SH2: Rotor angle with reference to reference machine angle in deg
TSB: Rotor angle with reference to reference machine angle in deg

FIGURE 6.17
Changing the unstable network shown in Figure 6.16, by adding PSS to bus 2.

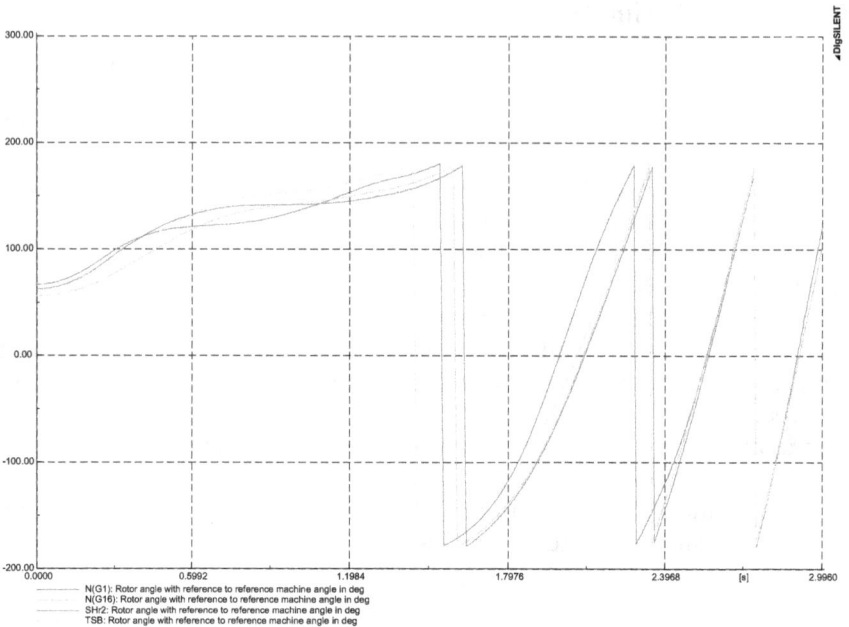

FIGURE 6.18
Changing the unstable network shown in Figure 6.16, by adding PSS to bus 4.

6.8 Two-Choice Questions (Yes/No)

1. PSS cannot enhance both small and large signal stability.
2. The network frequency or rotor speed is the traditional PSS input.
3. The power signal can be used in place of the speed signal to prevent PSS interaction.
4. The exciter needs to be connected to the PSS signal.
5. The ΔPC input in the PSS design block diagram is derived from optimal power flow.
6. The ΔV_{ref} in the PSS design block diagram is derived from reactive power control.
7. Reducing low-frequency oscillations is the primary PSS function in the power system.
8. Low-frequency oscillations are caused by connecting two distinct networks—one serving as a producer and the other as a consumer—with a single line.

9. The reset block is responsible for the PSS operating during regular oscillations.

10. PSS interference cannot be eliminated by using power signals rather than speed.

11. The inertia technique and the area division technique are reduction methods.

12. The model change technique and the modal approach are the reduction methods.

13. The coherency method models machines that oscillate in unison using a single machine.

14. The intensity of subsynchronous oscillations is reduced by changing the relevant capacitor.

15. The cause of SSR is that low-pressure and high-pressure turbines, exciter, and generator oscillate relative to each other.

16. Applying a small shock to the generator while it is operating at a low speed is the most effective, affordable, and secure method of identifying generator modes.

17. It is not possible to identify the generator's frequency modes by switching the network capacitors.

18. Frequency filtering and rotor damping can be used to remove oscillating modes.

19. The short circuit of the capacitor during resonance, and exciter-based generator control can be used to remove oscillating modes.

20. The ability of a power system to sustain synchronism in the face of minor perturbations is known as small-signal stability.

6.8.1 Answers to Two-Choice Questions

1, 9, 10, 14, 17: No
 Others: Yes

6.9 Appendix 1, Proof of Initial Conditions

As previously stated, the dynamic system's initial conditions $(V_q^0, V_d^0, I_q^0, I_d^0, \delta^0, E_0')$ must be established before the PSS parameters can be calculated. This appendix provides proof of identifying the network's initial conditions.

$$\hat{V}_t^0 = (V_q^0 + jV_d^0)e^{j\delta^0}, \quad \hat{I}_a^0 = (I_q^0 + jI_d^0)e^{j\delta^0}, \quad \left(V_t^0\right)^2 = \left(V_q^0\right)^2 + \left(V_d^0\right)^2 \tag{6.4}$$

$$S_e^0 = P_e^0 + jQ_e^0 = \hat{V}_t^0\left(\hat{I}_a^0\right)^* = V_q^0 I_q^0 + V_d^0 I_d^0 + j(V_d^0 I_q^0 - V_q^0 I_d^0) \tag{6.5}$$

With the above equations, the following five equations and five unknowns can be summarized.

1. $P_e^0 = V_q^0 I_q^0 + V_d^0 I_d^0$ $\hfill(6.6)$
2. $Q_e^0 = V_d^0 I_q^0 - V_q^0 I_d^0$ $\hfill(6.7)$
3. From generator equation : $V_q^0 = E_0' - r I_q^0 + x_d' I_d^0$ $\hfill(6.8)$
4. From generator equation : $V_d^0 = -x_q I_q^0 - r I_d^0$ $\hfill(6.9)$
5. $\left(V_t^0\right)^2 = \left(V_q^0\right)^2 + \left(V_d^0\right)^2$ $\hfill(6.10)$

Then:

6.

$$\left((\text{equation } 6.6)\times V_q^0\right)+\left((\text{equation } 6.7)\times V_d^0\right) \Rightarrow P_e^0 V_q^0 + Q_e^0 V_d^0 = \left(V_t^0\right)^2 I_q^0$$

$$\tag{6.11}$$

7.

$$\left((\text{equation } 6.6)\times V_d^0\right)+\left((\text{equation } 6.7)\times(-V_q^0)\right) \Rightarrow P_e^0 V_d^0 - Q_e^0 V_q^0 = \left(V_t^0\right)^2 I_d^0$$

$$\tag{6.12}$$

8. From equations (6.9), (6.11), and (6.12):

$$V_d^0 = -x_q\left(\frac{P_e^0 V_q^0 + Q_e^0 V_d^0}{\left(V_t^0\right)^2}\right) - r\left(\frac{P_e^0 V_d^0 - Q_e^0 V_q^0}{\left(V_t^0\right)^2}\right) \Rightarrow$$

$$\left(1 + \frac{x_q Q_e^0 + rP_e^0}{\left(V_t^0\right)^2}\right)V_d^0 = \left(\frac{-x_q P_e^0 + rQ_e^0}{\left(V_t^0\right)^2}\right)V_q^0 \Rightarrow \tag{6.13}$$

$$V_d^0 = \left(\frac{-x_q P_e^0 + rQ_e^0}{\left(V_t^0\right)^2 + x_q Q_e^0 + rP_e^0}\right)V_q^0 = A_{dq} V_q^0$$

That:

$$A_{dq} \triangleq \left(\frac{-x_q P_e^0 + rQ_e^0}{\left(V_t^0\right)^2 + x_q Q_e^0 + rP_e^0}\right)$$

9. From equations (6.10) and (6.13):

$$\left(V_t^0\right)^2 = \left(V_q^0\right)^2 + \left(A_{dq}\, V_d^0\right)^2 \rightarrow V_q^0 = \frac{\pm V_t^0}{\sqrt{1+A_{dq}^2}} \;,\; V_d^0 = A_{dq}\, V_q^0 \qquad (6.14)$$

10. From equations (6.11) and (6.12):

$$I_q^0 = \left(\frac{P_e^0\, V_q^0 + Q_e^0\, V_d^0}{\left(V_t^0\right)^2}\right),\; I_d^0 = \left(\frac{P_e^0\, V_d^0 - Q_e^0\, V_q^0}{\left(V_t^0\right)^2}\right) \qquad (6.15)$$

11. From equation (6.8):

$$E_0' = V_q^0 + r\, I_q^0 - x_d'\, I_d^0 \qquad (6.16)$$

In short, to determine the six dynamic system's initial conditions, equations (6.13) to (6.16) should be solved.

6.10 Appendix 2, The Effect of Exciter, Governor, and PSS in DIgSILENT

In this section, it is assumed that the reader has read the introductory information about the DIgSILENT PowerFactory and the governor and exciter from the previous chapters. If you are not interested or DIgSILENT PowerFactory is not available, you can ignore this section. Therefore, in the following, you will learn how to add exciter, governor, and PSS to the system in the software. In short, the governor is used to regulate the power, the exciter is used to regulate the voltage, and the PSS is used to stabilize the generator. In order to be able to follow this section, you must download the file (Chapter6-1.pfd) or (Chapter6-2.pfd) from the attached files at the end of the book. Import the network and follow the rest of this section.

To analyze this software comprehensively, the said methods are provided for both new (Version 2021 and after) and old (Version 15.1) editions. Additionally, a comparison between the two versions is conducted to identify any differences between the two.

6.10.1 First Part, Adding Governor and Turbine, Frequency Control

6.10.1.1 Version 15.1

In the opened folder, right-click on the G5 generator and click "Define" and look for "Governor and Turbine (gov)". Figure 6.19 shows how to copy and paste a "governor-turbine" into your library from the main library. In the left window, under "Database", select "Library" and then right-click on "pcu_HYGOV" as shown in Figure 6.19. After clicking "Copy," drag the

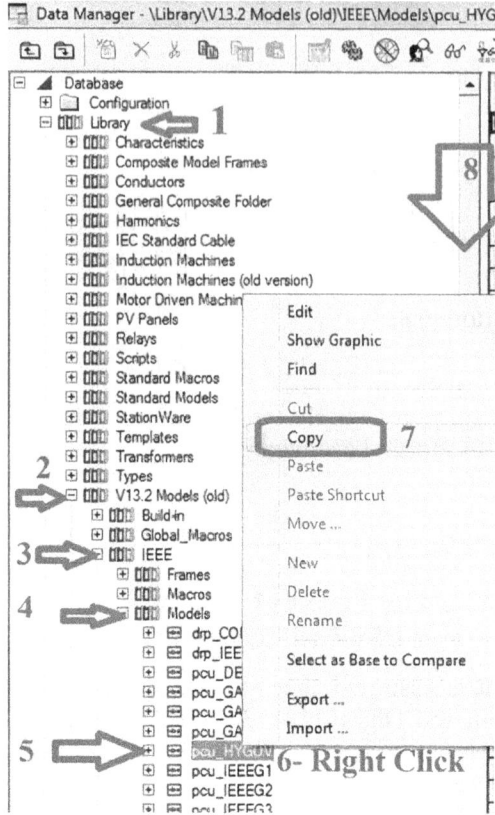

FIGURE 6.19
Copying the governor from the main library inside the studied network (15.1).

8 key down to reach Figure 6.20. After completing the activities in Figure 6.19, you can paste the file into your grid by right-clicking on the right side of the figure in the library (ticks 1, 2, and 3 in Figure 6.20). Now select "pcu_HYGOV" and click OK. The G5 generator must then be connected to the desired governor after the above has been correctly implemented. To be sure, proceed as shown in Figure 6.21. If it doesn't match, repeat 6.10.1.1.

6.10.1.2 Version 2021 and After

In the opened folder, right-click on the G5 generator and click "Define" and look for "Governor and Turbine (gov)". Figure 6.22 shows how to copy and paste a governor-turbine into your library from the main library. Right-click on the main library "DIgSILENT Library" in the window on the left, under "Database" (number 1, Figure 6.22), and select "gov_HYGOV". Using the 7 key, drag down until you reach Figure 6.23 after clicking "Copy".

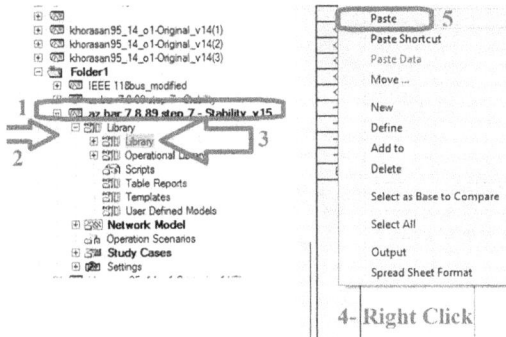

FIGURE 6.20
Pasting the governor to the studied network (15.1).

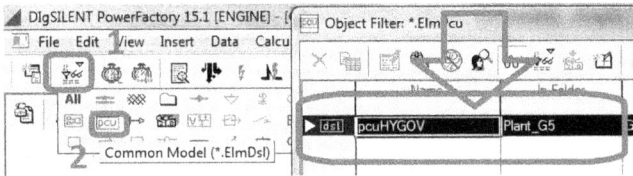

FIGURE 6.21
Checking the governor connection to the G5 generator (15.1).

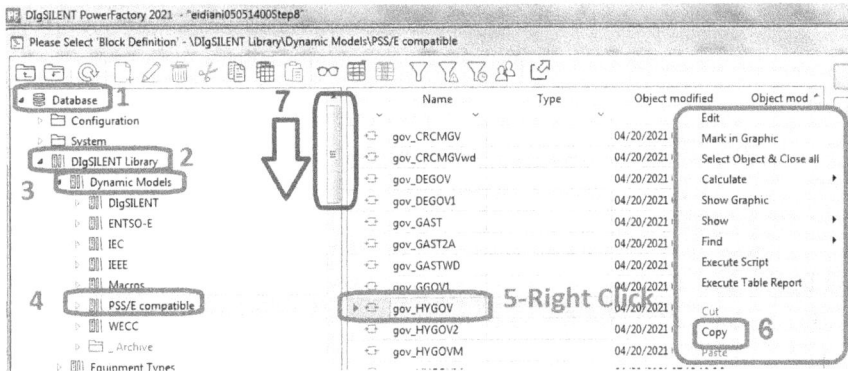

FIGURE 6.22
Copying the governor from the main library in the studied network (2021 edition and after).

Figure 6.23 shows how to click your network in the library, right-click (number 2) on the right side of the Figure, and then paste the file. Now select "gov_HYGOV" and click OK.

Following the correct implementation of the above, the desired governor must be connected to the G5 generator. To be sure, proceed as shown in Figure 6.24. If it does not match, repeat 6.10.1.2.

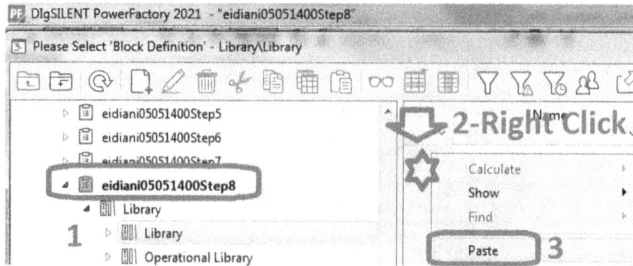

FIGURE 6.23
Pasting the governor to the studied network (2021).

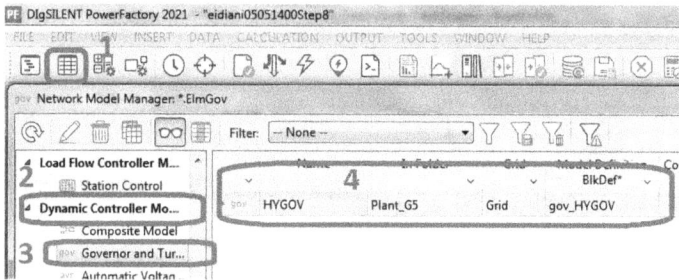

FIGURE 6.24
Checking the connection of the governor to the generator G5 (2021).

6.10.2 The Second Part, Adding AVR, Voltage Control

6.10.2.1 Version 15.1

Now again, as in "Section 6.10.1.1", right-click on the G5 generator and press the "**Define**" key, and in the opened folder, look for "**Automatic Voltage Regulator** (avr)". Now you need to go to the main library from the window that appears and copy and paste an AVR into your library. As in the previous step, this time select "vco_IEEET1", right-click, copy, and paste your network file into the library. Now select "vco_IEEET1" and click OK. Note: After pasting (Paste) AVR, be sure to tick AVR and then click OK (Figure 6.25), otherwise you will have to repeat this step.

6.10.2.2 Version 2021 and After

See Figure 6.26. Now again, as in "Section 6.10.1.2", right-click on the G5 generator and press the "Define" key, and in the opened folder, look for "Automatic Voltage Regulator (avr)". Now you need to go to the main library from the window that appears and copy and paste an AVR in your library. As in the previous step, this time select "avr_IEEET1", right-click, copy, and paste into your network file library (like 6.10.1.2). Now select "avr_IEEET1" and click OK.

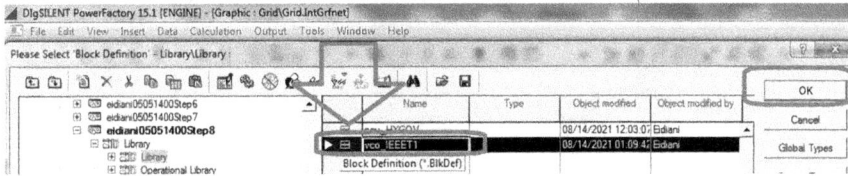

FIGURE 6.25
A very important point in pasting AVR (15.1).

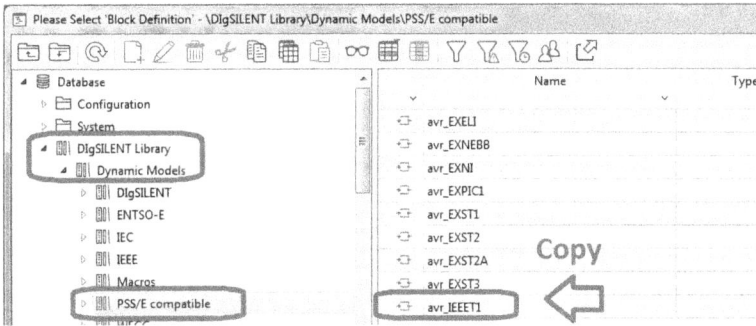

FIGURE 6.26
Choosing the fit AVR for the G5 generator (2021 edition).

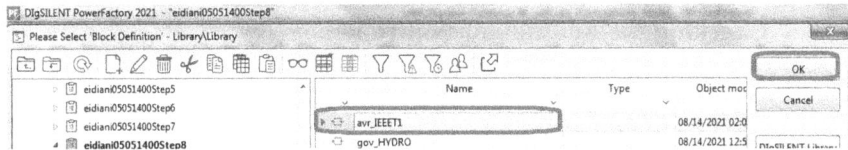

FIGURE 6.27
A very important point in pasting AVR (2021).

Note, after pasting the AVR, be sure to tick "AVR" and then click OK (Figure 6.27), otherwise you will have to repeat this step.

6.10.3 The Third Part, Adding PSS

6.10.3.1 Version 15.1

See Figure 6.28. Proceed exactly as in the previous two sections. On the G5 generator, right-click and press the "Define" key, and in the opened folder, select "Power System Stabilizer (pss)". As in Figure 6.19.

Select and paste "pss_STAB1" into your library. Click once. Figure 6.28 shows PSS parameters. Next, you should check the effect of these parameters on the network's transient stability by changing the PSS parameters.

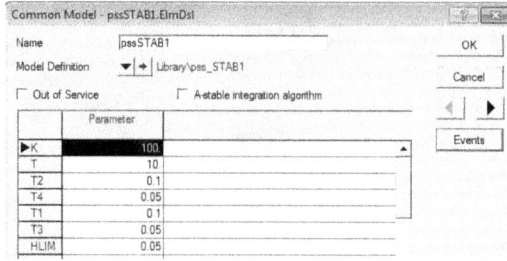

FIGURE 6.28
PSS parameters (15.1).

6.10.3.2 Version 2021 and After

Proceed exactly as in the previous two sections. Right-click on the G5 generator and press the "Define" key, and in the opened folder, select "Power System Stabilizer (pss)".

Select and paste "pss_STAB1" into your library. Click once. Figure 6.29 shows PSS parameters. Next, you should check the effect of these parameters on the network's transient stability by changing the PSS parameters.

Note: There are differences between the parameters "pss_STAB1" in versions 15.1 and 2021.

6.10.4 Checking the Addition of Controllers

6.10.4.1 Version 15.1

You can do this in two ways.

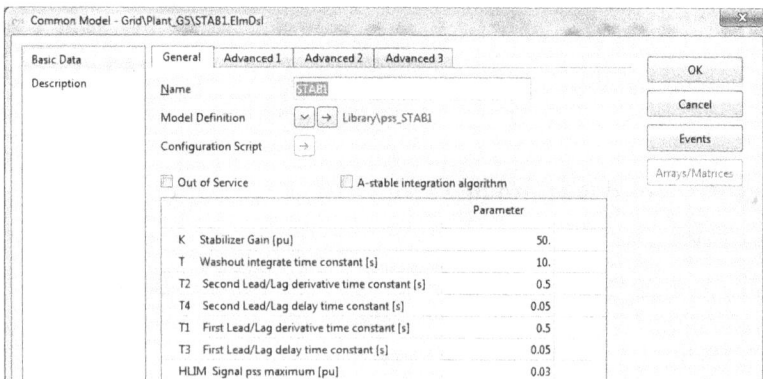

FIGURE 6.29
PSS parameters (2021).

6.10.4.1.1 Method 1

It is known from the "open data manager" if the controllers have entered your library from the main library or not. As shown in Figure 6.30, if these controllers are not connected, repeat the steps of adding controllers as in "Sections 6.10.1–6.10.3".

6.10.4.1.2 Method 2

As in Figure 6.31, four new icons should be added to the network's general information. Open them. There must be a controller in each of the controllers. If there were more, those with parentheses; delete them. You will need to repeat the steps if none of these icons appear. Repeat pcu "Section 6.10.1", vcu Section "6.10.2" and PSS "Section 6.10.3".

   Right-click on G5. Select the "Define" key and the desired controller. Repeat as in Figure 6.32.

6.10.4.2 Version 2021 and After

You can do this in two ways.

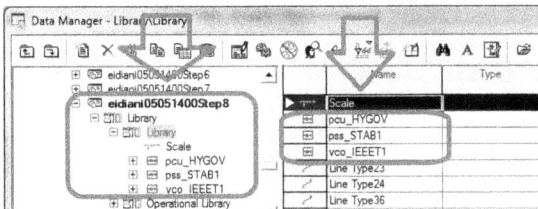

FIGURE 6.30
Checking the addition of generator controllers (version 15.1) method 1.

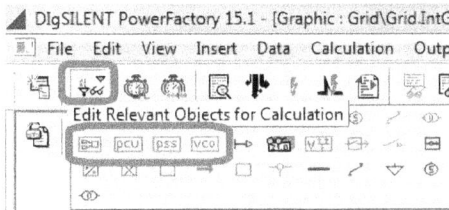

FIGURE 6.31
Checking the addition of generator controllers (version 15.1) method 2.

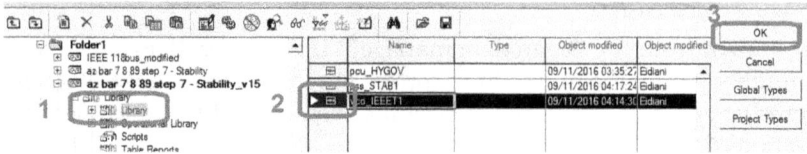

FIGURE 6.32
Checking the addition of generator controllers in a different way (version 15.1).

6.10.4.2.1 Method 1

It is known from "Data Manager" (number 1 in Figure 6.33) whether the controllers have entered your library from the main library or not. As shown in Figure 6.33, if these controllers are not connected, repeat the steps to add controllers as in "Sections 6.10.1–6.10.3".

6.10.4.2.2 Method 2

As in Figure 6.34, four new icons should be added to the network's general information. Open them. There must be a controller in each of the controllers. If there were more, those with parentheses; delete them. You will need to repeat the steps if none of these icons appear. Repeat pcu "Section 6.10.1", vcu "Section 6.10.2" and PSS "Section 6.10.3".

6.10.4.3 Composite Model

With the "**Composite Model**" you can take all controllers out of service. Each controller can be removed separately from the circuit in each section. These sections also contain information about each controller. Figure 6.35a for version 15.1 and Figure 6.35b for version 2021.

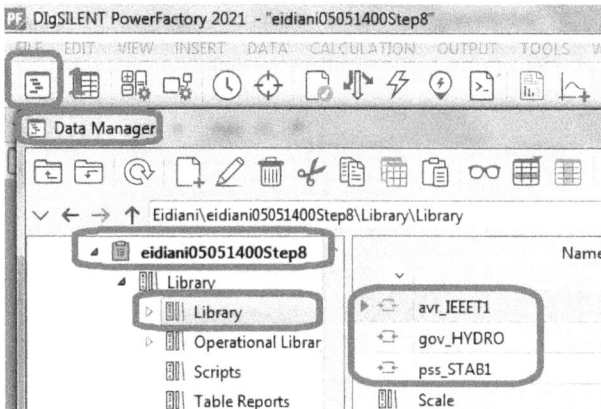

FIGURE 6.33
Checking the addition of generator controllers (2021 edition).

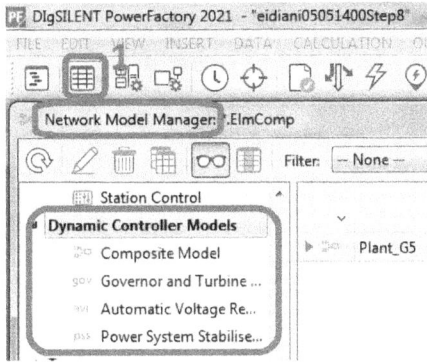

FIGURE 6.34
Checking the addition of generator controllers in a different way (2021 version).

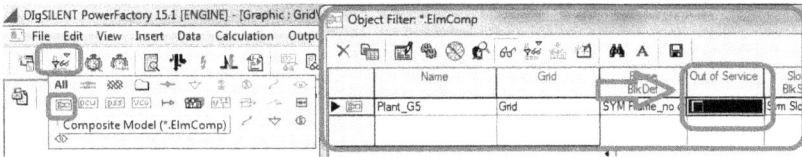

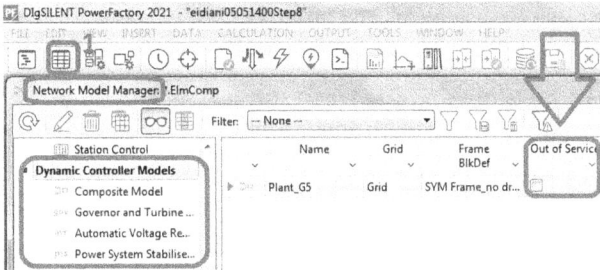

FIGURE 6.35
(a) How to out-of-service controller (version 15.17). (b) How to out-of-service controller (version 2021).

6.11 Summary

This chapter provides a clear overview of the functionality and design of PSS in electric power systems, including a step-by-step analysis of PSS design for a generator connected to an infinite bus. It offers a consolidated MATLAB program for PSS design formulas and showcases PSS performance through simulations in DIgSILENT PowerFactory on both a 9-bus power grid and a larger real grid. Additionally, the appendix includes guidance on software usage and suggestions for further research, with references for more in-depth study.

References

1. Kundur, P., *Power System Stability & Control*, McGraw-Hill, 1994.
2. Padiyar, K.R., *Power System Dynamics, Stability and Control*, 2nd edition, Indian Institute of Science, SSP BS Publications, Bangalore, 2008.
3. Machowski, J., Bialek, J.W., Bumby, J.R., *Power System Dynamics Stability and Control*, John Wiley & Sons, 2008.
4. Wang, X.F., Song, Y., Irving, M., *Modern Power Systems Analysis*, Springer, 2008.
5. Shubhanga, K.N., *Power System Analysis: A Dynamic Perspective*, 1st edition, Pearson Education, 2018.
6. Gomez-Exposito, A., Conejo, A.J., Canizares, C., *Electric Energy Systems Analysis and Operation*, 2nd edition, CRC Press, 2018.
7. Mondal, D., Chakrabarti, A., Sengupta, A., *Power System Small Signal Stability Analysis and Control*, 2nd edition, Elsevier Science and Technology Publication, 2020.
8. Baydokhty, M.E., Eidiani, M., Zeynal, H., Torkamani, H., Mortazavi, H., "Efficient generator tripping approach with minimum generation curtailment based on Fuzzy system rotor angle prediction," *Przeglad Elektrotechniczny*, 2012, 88(9A), pp. 266–271.
9. Eidiani, M., Kargar, M., Zeynal, H. "Interactive use of D-STATCOM and storage resource to maintain microgrid stability for commercial systems," In: Sivaraman, P., Sharmeela, C., Sanjeevikumar, P. (eds) *Microgrids for Commercial Systems*. Springer, Cham, 2024, pp. 241–270. https://doi.org/10.1002/9781394167319.ch10
10. Rouzbehi, K., Zhang, W., Candela, J. I., Luna, A., Rodriguez, P., "Unified reference controller for flexible primary control and inertia sharing in multi-terminal voltage source converter-HVDC grids," *IET Generation, Transmission & Distribution*, vol. 11, no. 3, pp. 750–758, 2017.
11. Chamorro, H. R., Sevilla, F. R. S., Gonzalez-Longatt, F., Rouzbehi, K., Chavez, H., "Innovative primary frequency control in low-inertia power systems based on wide-area RoCoF sharing," *IET Energy Systems Integration*, vol. 2, no. 2, pp. 151–160, 2020.
12. Zhang, W., Rouzbehi, K., Luna, A., Gharehpetian, G. B., Rodriguez, P., "Multi-terminal HVDC grids with inertia mimicry capability," *IET Renewable Power Generation*, vol. 10, no. 6, pp. 752–760, 2016.
13. Rakhshani, E., Luna, A., Rouzbehi, K., Rodriguez, P., Etxeberria-Otadui, I., "Effect of VSC-HVDC on load frequency control in multi-area power system," *2012 IEEE Energy Conversion Congress and Exposition* (ECCE), pp. 4432–4436, 2012.
14. Rouzbehi, K., Zhu, J., Zhang, W., Gharehpetian, G. B., Luna, A., Rodriguez, P., "Generalized voltage droop control with inertia mimicry capability-step towards automation of multi-terminal HVDC grids," *2015 International Conference on Renewable Energy Research and Applications*, 2015.
15. Zhang, W., Rouzbehi, K., Candela, J. I., Luna, A., Gharehpetian, G. B., Rodriguez, P., "Autonomous inertia-sharing control of multi-terminal VSC-HVDC grids," *2016 IEEE Power and Energy Society General Meeting* (PESGM), pp. 1–5, 2016.

7

Asynchronous Motors

7.1 Introduction

The three-phase symmetrical asynchronous or induction motor is the most widely used motor among all motor types. The reason is that this motor is cost-effective due to its long lifespan.

Symmetrical q-phase motors work similarly to three-phase symmetrical asynchronous motors, and they are much simpler to analyze than asymmetric or single-phase motors. In these motors, the stator windings are quite similar. A uniform and circular magnetic field is produced as a result, and their rotor has a symmetrical structure. This ensures a uniform distribution of induced currents in the rotor. These motors work with a balanced three-phase power supply, where all three phases have the same voltage and have a phase difference of 120°. Symmetrical motors have higher efficiency, less vibration and noise, simpler analysis, and stable performance.

Asynchronous machines are widely used in pumps, fans, compressors, conveyors, various industrial machines, generators in hydro and wind power plants, and traction motors in electric locomotives and subways. In addition, it is used in some industrial refrigerators, washing machines, and air conditioners, although single-phase motors are more common for domestic applications.

Without altering its internal structure, a standard induction motor can be transformed into an induction generator. To accomplish this, the induction motor's rotor must be mechanically moved by an external driver at a speed greater than the synchronous speed; in this scenario, the machine's slip becomes negative, and the induction motor transforms into an induction generator. Due to their low adjustment requirements and ease of control, induction generators are commonly found in wind farm and small hydropower plants. Chapter 13 discusses wind power plant analysis and how to build a wind farm using an asynchronous generator.

There are three sections to this chapter. The first section provides a brief overview of the classical analysis of the asynchronous motor and generator. The asynchronous motor is examined in the second section by using DIgSILENT PowerFactory software to simulate its operation, version 15.1

(v15.1) and version 2021 and later (v2021). The third section goes over how to use PowerFactory software for asynchronous machine analysis.

For more detailed work and further work, you can refer to the references [1–12] at the end of this chapter.

7.2 The Classical Analysis of the Asynchronous Machines

This section is appropriate for people who wish to quickly review the relationships and fundamental ideas of asynchronous machines but have forgotten them. The reader can proceed with the asynchronous machine analysis from the third section if they are already familiar with the fundamental ideas.

7.2.1 Rotating Magnetic Field Theory

Several symmetrical (similar) windings in the stator and a rotor with short circuit windings are characteristics of induction or asynchronous motors. Rotating transformers are another name for induction motors. Except for the fact that the rotor wires are shorted and cause it to rotate, the electric motor and the transformer both operate on the principle of induction, which is the basis for this comparison.

A magnetic field that seems to revolve in a certain direction is called a rotating magnetic field. Many electrical devices, particularly induction motors, depend on this phenomenon to function. A multiphase current source can be used to create a rotating magnetic field. A spinning field is created from two comparable coils with a 90-degree spatial difference in a symmetrical two-phase motor by two currents of equal magnitude and a 90-degree angle difference. Three identical coils with a spatial difference of ±120 degrees, three currents of equal size, and an angle difference of ±120 degrees are used to create a rotating field in a symmetrical three-phase motor. In symmetric motors, only a left-handed or right-handed field is produced due to the spatial and electric angle similarity. Around the core, this magnetic field seems to be constantly rotating. The number of poles in the stator coil and the frequency of the AC source control how quickly this magnetic field rotates.

This spinning field induces a voltage in the rotor coils after it passes through them, and the short circuit of the rotor coils results in the creation of current. The asynchronous motor rotates and creates torque when a force is applied to the electric current-carrying wire in accordance with physical rules.

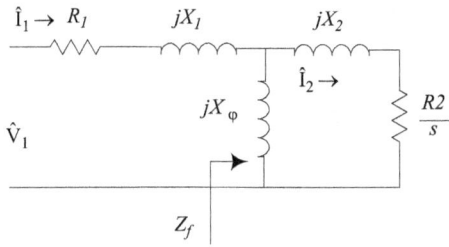

FIGURE 7.1
One-phase equivalent circuit for a *q*-phase symmetrical balanced asynchronous machine.

In the following, the equivalent circuit of the asynchronous motor is examined.

7.2.2 Equivalent Circuit Model of Asynchronous Machines

Figure 7.1 illustrates a famous one-phase equivalent circuit for a *q*-phase symmetrical balanced asynchronous machine.
That:

Resistance of one phase of the stator winding: R_1 or R_s

Reactance of one phase of the stator winding: X_1 or X_s

Resistance of one phase of the rotor winding (referred to as the stator side): R_2 or R_r

Reactance of one phase of the rotor winding (referred to as the stator side): X_2 or X_r

Magnetizing reactance: X_m or X_φ

One-phase voltage phasor: $\hat{V}_1$

The current of one phase of the stator: $\hat{I}_1$

The current of one phase of the rotor (referred to as the stator side): $\hat{I}_2$

$$\text{Slip: } s = \frac{n_s - n_r}{n_s} = \frac{\omega_s - \omega_r}{\omega_s} \Rightarrow \begin{cases} \omega_r = (1-s)\omega_s \\ n_r = (1-s)n_s \end{cases} \tag{7.1}$$

Stator field speed (revolutions per minute or rpm):

$$n_s = \frac{120 f_s}{P} \tag{7.2}$$

Stator field speed $\left(\text{radian per second}\right): \omega_s = \frac{4\pi f_s}{P} \tag{7.3}$

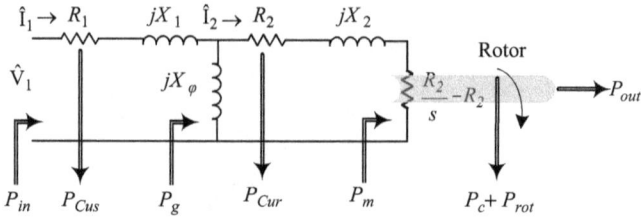

FIGURE 7.2
Complete one-phase equivalent circuit for a q-phase symmetrical balanced asynchronous machine.

Synchronous frequency: f_s

The number of poles: $Pole$

The frequency of the voltage induced in the rotor winding: $f_r = s\,f_s$ (7.4)

Instead of Figure 7.1, Figure 7.2 is used if power transfer details are concerned.

The equations of the q-phase asynchronous machine are as follows.

$$\text{Input active power: } P_{in} = q\,V_1\,I_1\cos(\varphi) \tag{7.5}$$

$$\text{Stator copper loss: } P_{Cus} = q\,R_1\,I_1^2 \tag{7.6}$$

$$\text{Air-gap power: } P_g = P_{in} - P_{Cus} = q\left(\frac{R_2}{s}\right)I_2^2 \tag{7.7}$$

$$\text{Rotor copper loss: } P_{Cur} = q\,R_2\,I_2^2 = s\,P_g \tag{7.8}$$

$$\text{Mechanical power: } P_m = P_g - P_{Cur} = (1-s)P_g \tag{7.9}$$

$$\text{Mechanical torque: } T_m = \frac{P_m}{\omega_r} = \frac{P_g}{\omega_s} \tag{7.10}$$

Fixed core loss: P_c

Fixed rotational loss (wind and friction): P_{rot}

$$\text{Output power: } P_{out} = P_m - P_c - P_{rot} = P_{in} - \overbrace{(P_{Cus} + P_{Cur} + P_c + P_{rot})}^{P_{Loss}} \tag{7.11}$$

$$\text{Total loss: } P_{Loss} = P_{Cus} + P_{Cur} + P_c + P_{rot} \tag{7.12}$$

$$\text{Output torque: } T_{out} = \frac{P_{out}}{\omega_r} \tag{7.13}$$

$$\text{Efficiency: } \eta = \frac{P_{out}}{P_{in}} = \frac{P_{out}}{P_{out} + P_{Loss}} = \frac{P_{in} - P_{Loss}}{P_{in}} \qquad (7.14)$$

As in Figure 7.1, we must use the Thévenin impedance on the rotor side Z_f to solve the equivalent circuit of Figure 7.1 or 7.2.

$$\text{Thevenin impedance on the rotor side: } Z_f = R_f + jX_f = (jX_\varphi) \| \left(\frac{R_2}{s} + jX_2 \right) \qquad (7.15)$$

The rest of the variables can now be calculated as follows.

$$\text{Input impedance: } Z_{in} = Z_1 + Z_f = (R_1 + jX_1) + (R_f + jX_f) \qquad (7.16)$$

$$\text{Input or stator current: } \hat{I}_1 = \frac{\hat{V}_1}{Z_{in}} \qquad (7.17)$$

$$\text{Input power factor: } \cos(\varphi) = \cos(\angle V_1 - \angle I_1) \qquad (7.18)$$

Furthermore, air-gap power equations (7.7) and (7.8) can be extended with one more relation.

$$P_g = q\,R_f\,I_1^2 = P_{in} - P_{Cus} = q\left(\frac{R_2}{s} \right)I_2^2 = \frac{P_{Cur}}{s} = \frac{P_m}{(1-s)} \qquad (7.19)$$

As a last point, the speed of the stator field can be arranged in two Tables. Using Tables 7.1 and 7.2, we calculate the frequency, number of poles, stator field speed, and slip based on the known rotor speed. Typically, the stator's rotating field speed is marginally slower than the rotor's. Knowing the rotor speed, the other parameters can be guessed.

TABLE 7.1

Stator Rotating Field speed (rpm and rad/s), f_s=50 Hz, from equations (7.2) and (7.3)

Pole	2	4	6	8	10	12	14	16	18
n_s	3000	1500	1000	750	600	500	428.57	375	333.33
ω_s	100 π	50 π	33.33 π	25 π	20 π	17.67 π	17.29 π	12.5 π	11.11 π

TABLE 7.2

Stator Rotating Field Speed (rpm and rad/s), f_s=60 Hz, from equations (7.2) and (7.3)

Pole	2	4	6	8	10	12	14	16	18
n_s	3600	1800	1200	900	720	600	517.29	450	400
ω_s	120 π	60 π	40 π	30 π	24 π	20 π	17.14 π	15 π	13.33 π

Example 7.1

If $\omega_r = 39\,\pi$ the closest stator field speed to this rotor speed is $(\omega_s = 40\,\pi)$ in Table 7.2. Then we have: $n_s = 1200$, $Pole = 4$ and $f_s = 60\,\mathrm{Hz}$ and $s = 0.025$

7.2.3 Maximum Torque

According to Figure 7.3, the Thévenin of the stator side should be calculated to determine the maximum torque.

Thevenin impedance and voltage can be calculated as follows.

Thevenin impedance on the stator side: $Z_e = R_e + jX_e = (jX_\varphi) \,||\, (R_1 + jX_1)$

$$\text{(7.20)}$$

Thevenin voltage on the stator side: $\hat{V}_e = \dfrac{jX_\varphi}{R_1 + j(X_\varphi + X_1)}(\hat{V}_1)$ (7.21)

Now we can calculate the rotor current.

$$\hat{I}_2 = \frac{\hat{V}_e}{Z_e + Z_2} = \frac{\hat{V}_e}{(R_e + jX_e) + \left(\dfrac{R_2}{s} + jX_2\right)} \tag{7.22}$$

From equations (7.7) and (7.10), we have:

$$T_m = \frac{q}{\omega_s}\left(\frac{R_2}{s}\right)I_2^2 = \frac{q}{\omega_s}\left(\frac{R_2}{s}\right)\frac{V_e^2}{\left(R_e + \dfrac{R_2}{s}\right)^2 + (X_e + X_2)^2} \tag{7.23}$$

This equation; which is the famous torque-slip curve, will be discussed in the next section. The slip is obtained at maximum torque if the derivative of the slip becomes zero (The verification with the reader).

Slip at maximum torque: $S_{T\max} = \dfrac{R_2}{\sqrt{R_e^2 + (X_e + X_2)^2}}$ (7.24)

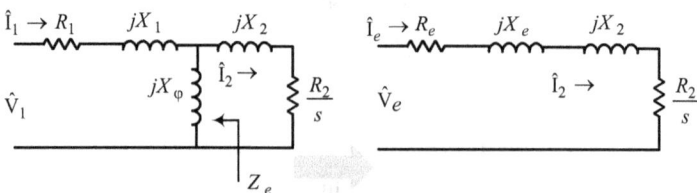

FIGURE 7.3
One-phase equivalent circuit to determine the maximum torque.

$$\text{Maximum torque:} T_{max} = \left(\frac{1}{\omega_s}\right)\frac{0.5(q)(V_e^2)}{R_e + \dfrac{R_2}{S_{T\,max}}} = \left(\frac{1}{\omega_s}\right)\frac{0.5(q)(V_e^2)}{R_e + \sqrt{R_e^2 + (X_e + X_2)^2}}$$

$$(7.25)$$

7.2.4 Two-Choice Questions (Yes/No)

1. Increasing the rotor resistance always increases the starting torque.
2. Increasing the rotor resistance always decreases the starting current.
3. The rotor resistance can only be increased.
4. A decrease in voltage will decrease the motor speed.
5. The asynchronous motor starting current depends on the load.
6. The starting time of the asynchronous motor does not depend on the load.
7. By adding suitable resistance to the rotor, the starting torque is increased.
8. The starting torque cannot be maximized by adding the appropriate resistance to the rotor.
9. A rotor's rotation direction can be changed by switching to two phases.
10. When one phase of the three-phase motor is interrupted, the motor starts in no-load mode.
11. The motor rotation direction depends on the axis direction of the stator and rotor.
12. In a balanced machine, the size of the current is equal, and the angle difference (±120).
13. The speed of the rotating field is a function of frequency and the number of poles.
14. The slip range of an induction motor is between zero and minus one.
15. In an induction motor, there must be zero slip to generate torque.
16. Slip is a function of load in an induction motor.

7.2.4.1 Key Answers to Two-Choice Questions

Yes	2–4, 7, 9, 10, 12, 13, 16
No	1, 5, 6, 8, 11, 14, 15

7.2.5 Descriptive Questions for the Symmetrical Asynchronous Machines

7-1 If (Pole=4) and (f=60 Hz), determine the frequency of the voltage induced in the rotor of an asynchronous machine.

    a. The rotor is not moving.

    b. At 1800 rpm, the rotor rotates in the direction of the stator field.

    c. At 1800 rpm, the rotor rotates in the opposite direction to the stator field.

         Difficulty level ● Easy ○ Normal ○ Hard

7-2 If the rotor speed of an asynchronous motor is (ω_r=49 π rad/s), find:

    a. f_s, Pole, ω_s, n_s and s.

    b. Slip for the third harmonic

    c. Slip for the fifth harmonic

         Difficulty level ● Easy ○ Normal ○ Hard

7-3 The rotor in Figure 7.4 is connected to a 50 Hz, 2-pole three-phase positive balanced source. The rotor rotates in the direction of its rotating field at a speed of 1000 rpm. Find the following.

    a. The speed of the rotor field, relative to the rotor.

    b. The speed of the rotor field, relative to the stator.

    c. The frequency of the voltage induced in the stator.

        If the rotation direction of the rotor is reversed, find the values of (a), (b), and (c) again.

         Difficulty level ● Easy ○ Normal ○ Hard

7-4 A three-phase asynchronous motor (balanced, symmetrical), and has the following stator side characteristics:

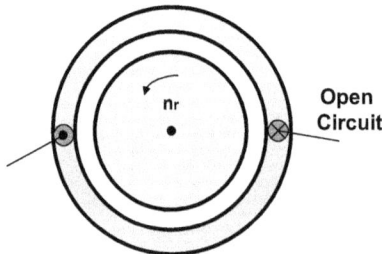

FIGURE 7.4
Figure question 7-3.

$$60 \text{ Hz, Y, 220 V, 7 kW, 6 Poles, } R_1 = R_2 = 0.3 \text{ } \Omega,$$

$$X_1 = 0.5 \text{ } \Omega, X_2 = 0.2 \text{ } \Omega, X_\varphi = 10 \text{ } \Omega$$

$$s = 0.03, P_{rot} + P_c = 300 \text{ W}$$

Find: ω_r, n_r, I_s, $\cos(\varphi)$, P_{in}, P_g, P_m, P_{out}, T_{out}, η, $s_{T\max}$, T_{max}
Difficulty level ● Easy ○ Normal ○ Hard

7-5 In **question 7-4**, If with the star-delta switch, the motor becomes delta. Calculate the speed of the rotor.
Difficulty level ● Easy ○ Normal ○ Hard

7-6 In response to **question 7-4**, we swap out the two phases to make the motor stop. How many times will the rotor and stator currents be multiplied in this scenario?
Difficulty level ● Easy ○ Normal ○ Hard

7-7 The starting current in **question 7-4** should match the rated current. Determine how much resistance has been added to the rotor winding.
Difficulty level ● Easy ○ Normal ○ Hard

7-8 Our goal in answering **question 7-4** is to maximize the initial torque. Determine how much resistance has been added to the rotor winding.
Difficulty level ● Easy ○ Normal ○ Hard

7-9 In **question 7-4**, determine the asynchronous generator's output power if the rotor speed reaches 2100 rpm.
Difficulty level ● Easy ○ Normal ○ Hard

7-10 Using the given specifications, find the asynchronous motor's input current

$$220 \text{ V, 8 kW, } R_1 = R_2 = 0.3 \text{ } \Omega, X_1 = X_2 = 0.5 \text{ } \Omega, X_\varphi = \infty, s = 0.03$$

a. 37.36   b. 12.27   c. 21.26   d. 20.99

7-11 Repeat **questions 7-10** with $(s = -0.03)$.

a. 13.03   b. 22.56   c. 37.36   d. 9.75

7-12 Using the following details, determine the asynchronous motor's speed (rpm).

$$50 \text{ Hz, 6 Pole, } T_{max} = 100 \text{ N.m, } S_{T\max} = 0.2, T_{Start} = 30 \text{ N.m, } T_n = 25 \text{ N.m}$$

a. 0   b. 1140   c. 950   d. 977.6

7-13 Using the following details, determine the rotor copper loss.

$$n_s = 1500 \text{ rpm, } n_r = 1200 \text{ rpm, } T_{max} = 90 \text{ N.m, } T_{out} = 60 \text{ N.m}$$

a. 600π b. 3000π c. 300π d. 6000π

7–14 Using the following details, determine the maximum torque.

220 V, 50 Hz, 6 Pole, $R_1 = R_2 = 0.2\ \Omega$, $X_1 = X_2 = 0.5\ \Omega$, $X_\varphi = 30\ \Omega$

   a. 186   b. 61.8   c. 19.4   d. 223

7–15 Which of the following relationships is correct?

   a. $T = \dfrac{2T_{max}}{\dfrac{s_{T\,max}}{s} + \dfrac{s}{s_{T\,max}}}$
   b. $T = \dfrac{2T_{max}}{\dfrac{s_{T\,max}}{s}}$
   c. $T = \dfrac{2T_{max}}{\dfrac{s}{s_{T\,max}}}$
   d. $T = \dfrac{s T_{max}}{s_{T\,max}}$

7–16 Which of the following relationships is correct?

   a. $s_{P\,max} = s_{T\,max}$
   b. $s_{P\,max} = 2 s_{T\,max}$
   c. $\dfrac{1}{s_{P\,max}} = \sqrt{1 + (\dfrac{1}{s_{T\,max}})^2}$

   d. $\dfrac{1}{s_{P\,max}} = 1 + \sqrt{1 + (\dfrac{1}{s_{T\,max}})^2}$

7.2.6 Descriptive Answers of the Symmetrical Asynchronous Machines

7-1 From equations (7.2), (7.3) and (7.4) we have: $f_r = s\ f_s$

$$n_s = \frac{120\ f_s}{P} = \frac{(120)(60)}{4} = 1800\ \text{rpm}$$

   a. $s = \dfrac{n_s - n_r}{n_s} = \dfrac{1800 - 0}{1800} = 1 \Rightarrow f_r = s\ f_s = 60\ \text{Hz}$

   b. $s = \dfrac{n_s - n_r}{n_s} = \dfrac{1800 - 1800}{1800} = 0 \Rightarrow f_r = s\ f_s = 0\ \text{Hz}$

   c. $s = \dfrac{n_s - n_r}{n_s} = \dfrac{1800 - (-1800)}{1800} = 2 \Rightarrow f_r = s\ f_s = 120\ \text{Hz}$

7-2 a. If $\omega_r = 49\ \pi$, the closest stator field speed to this rotor speed is ($\omega_s = 40\ \pi$) in Table 7.1. Then we have: $n_s = 1500$, Pole=4 and $f_s = 50\ \text{Hz}$ and $s = 0.02$

   b. and c. The question must be answered by the reader.

7-3 From equations (7.2) and (7.3) we have:

   a. $n_s = \dfrac{120\ f_s}{P} = \dfrac{(120)(50)}{2} = 3000\ \text{rpm}.$

   b. $3000 + 1000 = 4000\ \text{rpm}$

c. $\dfrac{4000}{60}\dfrac{\text{rpm}}{\text{m}} = 66.67$ Hz

d. 3000 rpm, e. 3000 − 1000 = 2000 rpm, f. $\dfrac{2000}{60}\dfrac{\text{rpm}}{\text{m}} = 33.33$ Hz

7-4 From equations (7.2) and (7.3) we have:

$$n_s = \frac{120\,f_s}{P} = \frac{(120)(60)}{6} = 1200 \text{ rpm}$$

$$\omega_s = \frac{4\pi\,f_s}{P} = \frac{4\pi(60)}{6} = 40\pi = 125.66\ \frac{\text{rad}}{\text{s}}$$

$$n_r = (1-s)n_s = (1-0.03)(1200) = 1164 \text{ rpm}$$

$$\omega_r = (1-s)\omega_s = (1-0.03)(125.66) = 121.89 \text{ rad/s}$$

$$\hat{V}_1 = \frac{220\angle 0}{\sqrt{3}} = 127.02\angle 0 \text{ V}$$

See Figure 7.5.

- **The first method**

    We simplify Figure 7.5 to Figure 7.6 and we have:

    1. (1-1) From equation (7.15), Thévenin impedance on the rotor side:

        $$Z_f = (j10)\,\|\,(10 + j0.2) = 4.901 + j5.001 = R_f + jX_f$$

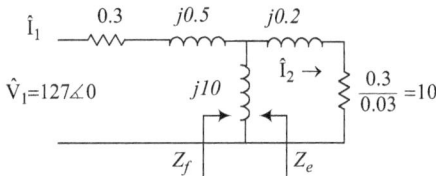

FIGURE 7.5
Figure 1 answer 7-4.

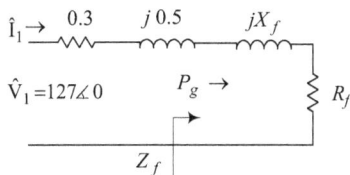

FIGURE 7.6
Figure 2 answer 7-4.

2. (1-2) Input impedance:
$Z_{in} = Z_1 + Z_f = (0.3 + j0.5) + (4.901 + j5.001) = 5.201 + j5.501$

$= 7.57 \angle 46.61°$

3. (1-3) Input or stator current:
$\hat{I}_1 = \dfrac{\hat{V}_1}{Z_{in}} = \dfrac{127.02 \angle 0}{7.57 \angle 46.61} = 16.78 \angle -46.61°$

4. (1-4) Input power factor: $\cos(\varphi) = \cos(46.61°) = 0.687$

5. (1-5) Input active power:
$P_{in} = 3V_1 I_1 \cos(\varphi) = 3(127.02)(16.78)(0.687) = 4393$ W

6. (1-6) Stator copper loss: $P_{Cus} = 3R_1 I_1^2 = 3(0.3)(16.78)^2 = 253$ W

7. (1-7) Air-gap power:
$P_g = 3R_f I_1^2 = 3(4.901)(16.78)^2 = 4140$ W $= P_{in} - P_{Cus}$

8. (1-8) Mechanical power:
$P_m = (1-s)P_g = (1-0.03)(4140) = 4016$ W

9. (1-9) Output power:
$P_{out} = P_m - P_c - P_{rot} = 4016 - 300 = 3716$ W

10. (1-10) Output torque: $T_{out} = \dfrac{P_{out}}{\omega_r} = \dfrac{3716}{121.89} = 30.49$ N.m

11. (1-11) Efficiency: $\eta = \dfrac{P_{out}}{P_{in}} = \dfrac{3716}{4393} = 0.8459 = 84.59\%$

12. (1-12) Thévenin impedance on the stator side:
$Z_e = (j10) \| (0.3 + j0.5) = 0.272 + j0.484 = R_e + jX_e$

13. (1-13) Thévenin voltage on the stator side:
$\hat{V}_e = \dfrac{j10}{0.3 + j(10 + 0.5)}(127.02 \angle 0) = 120.92 \angle 1.637°$

14. (1-14) Slip at maximum torque:
$S_{T\max} = \dfrac{R_2}{\sqrt{R_e^2 + (X_e + X_2)^2}} = \dfrac{0.3}{\sqrt{0.272^2 + (0.484 + 0.2)^2}} = 0.408$

15. (1-15) Maximum torque:
$T_{\max} = \left(\dfrac{1}{\omega_s}\right) \dfrac{0.5(3)(V_e^2)}{R_e + \dfrac{R_2}{S_{T\max}}} = \left(\dfrac{1}{125.66}\right) \dfrac{1.5(120.92)^2}{0.272 + \dfrac{0.3}{0.408}} = 173.27$ N.m

- **The second method**

  We simplify Figure 7.5 to Figure 7.7 and we have:

16. (2-1) = (1-12) Thévenin impedance on the stator side

17. (2-2) = (1-13) Thévenin voltage on the stator side

18. (2-3): Rotor current:
$\hat{I}_2 = \dfrac{\hat{V}_e}{Z_e + Z_2} = \dfrac{120.92 \angle 1.637°}{(0.272 + j0.484) + (10 + j0.2)} = 11.746 \angle -2.173°$

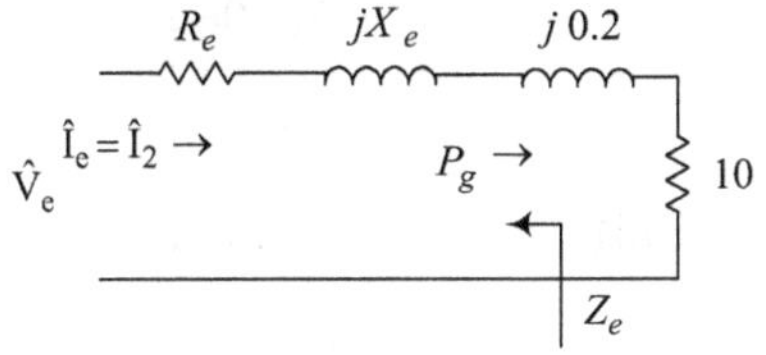

FIGURE 7.7
Figure 3 answer 7-4.

19. (2-4)=(1-3) Input or stator current: (KCL in Figure 7.5):

$$\hat{I}_2 = \hat{I}_1 \frac{jX_\varphi}{\dfrac{R_2}{s} + j(X_2 + X_\varphi)} \Rightarrow \hat{I}_1 = \frac{(10 + j10.2)(11.746\angle - 2.173°)}{j10}$$

$$= 16.78\angle - 46.61°$$

20. (2-5)=(1-4) Input power factor
21. (2-6)=(1-5) Input active power
22. (2-7)=(1-6) Stator copper loss
23. (2-8)=(1-7) Air-gap power:

$$P_g = 3\left(\frac{R_2}{s}\right)I_2^2 = 3(10)(11.746)^2 = 4139 \text{ W} \approx 4140 \text{ W}$$

24. (2-9)=(1-8) Mechanical power
25. (2-10)=(1-9) Output power
26. (2-11)=(1-10) Output torque
27. (2-12)=(1-11) Efficiency
28. (2-13)=(1-14) Slip at the maximum torque
29. (2-14)=(1-15) Maximum torque

7-5 New voltage from Y to Δ: $V_e = 120.92\sqrt{3}$ and
$T_{out} = T_{Load} = $ Constant $= 30.49$
  We have from equations (7.11) and (7.13):

$$T_{out} = \frac{P_{out}}{\omega_r} = \frac{P_m - P_c - P_{rot}}{\omega_r} = \frac{(1-s)P_g - 300}{(1-s)\omega_s} = \frac{P_g}{\omega_s} - \frac{300}{(1-s)\omega_s}$$

And we have from equations (7.19) and (7.22):

$$P_g = 3\frac{R_2}{s}I_2^2 = 3\frac{R_2}{s}\frac{V_e^2}{\left(R_e + \dfrac{R_2}{s}\right)^2 + (X_e + X_2)^2}$$

Then:

$$\left(\frac{1}{40\pi}\right)(3)\left(\frac{0.3}{s}\right)\frac{(120.92\sqrt{3})^2}{\left(0.272 + \dfrac{0.3}{s}\right)^2 + (0.484 + 0.2)^2} - \frac{300}{(1-s)40\pi} = 30.49$$

By trial and error, we have: $s = 0.0096 \Rightarrow n_r = (1 - 0.0096)(1200)$
$= 1188.5$ rpm

7-6 We have the slip in braking mode: $s_b = 2 - s = 2 - 0.03 = 1.97$
From equation (7.15), Thévenin impedance on the rotor side:

$$Z_f = (j10) \, \| \left(\frac{0.3}{1.97} + j0.2 \right) = 0.146 + j0.198 = R_f + jX_f$$

From equation (7.17): Input or stator current:

$$\hat{I}_1 = \frac{\hat{V}_1}{Z_{\text{in}}} = \frac{127.02 \angle 0}{(0.3 + j0.5) + (0.146 + 0.198)} = 153.346 \angle -57.42°$$

From equation (7.22):

$$\hat{I}_2 = \frac{\hat{V}_e}{Z_e + Z_2} = \frac{120.92 \angle 1.637°}{(0.272 + j0.484) + \left(\dfrac{0.3}{1.97} + j0.2 \right)} = 150.229 \angle -56.55°$$

Then: Stator current ratio:

$$\frac{I_1(\text{braking})}{I_1(\text{rated})} = \frac{153.346}{16.78} = 9.139$$

Rotor current ratio:

$$\frac{I_2(\text{braking})}{I_2(\text{rated})} = \frac{150.229}{11.746} = 12.79$$

7-7 From equations (7.15), (7.16) and (7.17):

$$I_1^{\text{Rated}} = I_1^{\text{Start}} \Rightarrow \frac{R_2}{0.03} = \frac{R_2 + \Delta R_2}{1} \Rightarrow \frac{0.3}{0.03} = 0.3 + \Delta R_2 \Rightarrow \Delta R_2 = 9.7 \, \Omega$$

7-8 From equation (7.24):

$$S_{T\max} = 1 \Rightarrow S_{T\max} = \frac{R_2 + \Delta R_2}{\sqrt{R_e^2 + (X_e + X_2)^2}} = 1 \Rightarrow$$

$$\Rightarrow \frac{0.3 + \Delta R_2}{\sqrt{0.272^2 + (0.484 + 0.2)^2}} = 1 \Rightarrow \Delta R_2 = 0.436$$

7-9 From equation (7.1):

$$S^{\text{Gen}} = \frac{n_s - n_r}{n_s} = \frac{1200 - 2100}{1200} = -0.75$$

From equation (7.15): Thévenin impedance on the rotor side:

$$Z_f = (j10) \,\|\, \left(\frac{0.3}{-0.75} + j0.2 \right) = -0.384 + j0.211 = R_f + jX_f$$

From equation (7.17), input or stator current:

$$\hat{I}_1 = \frac{\hat{V}_1}{Z_{in}} = \frac{127.02 \angle 0}{(0.3 + j0.5) + (-0.384 + j0.211)} = 177.416 \angle -96.738°$$

From equation (7.5), input active power:

$$P_{in} = 3\,V_1\,I_1 \cos(\varphi) = 3(127.02)(177.416)(\cos(96.738)) = -7932 \text{ W}$$

Negative input power means output power.

7–10 Answer (b) is correct. (See Figure 7.8). We have:

$$\hat{I}_1 = \frac{\hat{V}_1}{Z_{in}} = \left(\frac{220}{\sqrt{3}} \right) \left(\frac{1 \angle 0}{(0.3 + j0.5) + (10 + j0.5)} \right) = 12.27 \angle -5.55°$$

7-11 Answer (a) is correct. (See Figure 7.9). We have:

$$\hat{I}_1 = \frac{\hat{V}_1}{Z_{in}} = \left(\frac{220}{\sqrt{3}} \right) \left(\frac{1 \angle 0}{(0.3 + j0.5) + (-10 + j0.5)} \right) = 13.03 \angle -174.1°$$

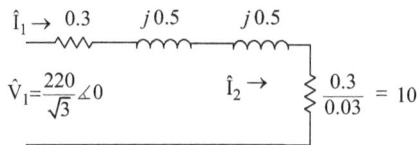

FIGURE 7.8
Figure answer 7-10.

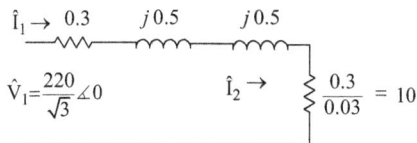

FIGURE 7.9
Figure answer 7-11.

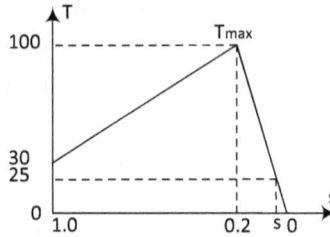

FIGURE 7.10
Figure answer 7-12.

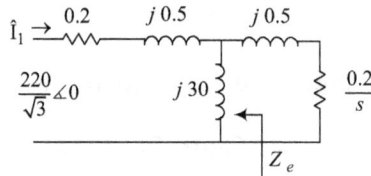

FIGURE 7.11
Figure answer 7-14.

7-12 Answer **(c)** is correct. See Figure 7.10. We have proportionally:

$$\frac{T_1}{T_2} = \frac{s_1}{s_2} \Rightarrow \frac{100}{25} = \frac{0.2}{s} \Rightarrow s = 0.05, \text{ and } n_s = \frac{120\,f_s}{P} = \frac{120(50)}{6} = 1000 \text{ rpm}$$

$$\Rightarrow n_r = (1-s)n_s = (1-0.05)(1000) = 950 \text{ rpm}$$

7-13 Answer **(a)** is correct. From equation (7.1):

$$s = \frac{n_s - n_r}{n_s} = \frac{1500 - 1200}{1500} = 0.2, \text{ and, } \omega_s = \frac{\pi}{30} n_s = 50\pi$$

$$P_g = T_{out}\omega_s = 60(50\pi) = 3000\pi \Rightarrow P_{Cur} = s\,P_g = 0.2(3000\pi) = 600\pi$$

7–14 Answer **(a)** is correct. (See Figure 7.11). We have:
Thévenin impedance on the stator side:

$$Z_e = R_e + jX_e = (j30)\,\|\,(0.2 + j0.5) = 0.193 + j0.493$$

Thévenin voltage on the stator side:

$$\hat{V}_e = \frac{j30}{0.2 + j(30.5)}\left(\frac{220}{\sqrt{3}} \angle 0\right) = 124.93 \angle -0.376°$$

$$S_{T\max} = \frac{R_2}{\sqrt{R_e^2 + (X_e + X_2)^2}} = \frac{0.2}{\sqrt{0.193^2 + (0.493 + 0.5)^2}} = 0.198$$

$$T_{\max} = \left(\frac{1}{\omega_s}\right)\frac{0.5(3)(V_e^2)}{R_e + \dfrac{R_2}{S_{T\max}}} = \left(\frac{1}{4\pi(50)}\right)\frac{1.5(124.93)^2}{0.193 + \dfrac{0.2}{0.198}} = 185.8 \text{ N.m}$$

7–15 Answer (a) is correct. Suppose that $(R_1 \simeq 0)$, and from (7.20) we have:

$$Z_e = R_e + jX_e = (jX_\varphi) \| (R_1 + jX_1) \Rightarrow R_e = 0$$

From (7.24):

$$\Rightarrow S_{T\max} = \frac{R_2}{\sqrt{R_e^2 + (X_e + X_2)^2}} = \frac{R_2}{X_e + X_2} \triangleq \frac{R_2}{X_T} \tag{7.26}$$

From (7.25):

$$T_{\max} = \frac{0.5(q)(V_e^2)}{\omega_s X_T} \tag{7.27}$$

From (7.23):

$$\Rightarrow T_m = T = \frac{q}{\omega_s}\left(\frac{R_2}{s}\right)\frac{V_e^2}{\left(\dfrac{R_2}{s}\right)^2 + (X_T)^2} \tag{7.28}$$

From $(7.28) \div (7.27)$, we have:

$$\frac{T}{T_{\max}} = \frac{\dfrac{q}{\omega_s}\left(\dfrac{R_2}{s}\right)\dfrac{V_e^2}{\left(\dfrac{R_2}{s}\right)^2 + (X_T)^2}}{\dfrac{0.5(q)(V_e^2)}{\omega_s X_T}} = \frac{2\left(\dfrac{R_2}{s}\right)X_T}{\left(\dfrac{R_2}{s}\right)^2 + (X_T)^2} \Rightarrow$$

$$\frac{T}{T_{\max}} = \frac{2}{\left(\dfrac{R_2}{s}\right)\dfrac{1}{X_T} + (X_T)\dfrac{s}{R_2}} \overset{\text{From (4.26)}}{=} \frac{2}{\dfrac{S_{T\max}}{s} + \dfrac{s}{S_{T\max}}}$$

7–16 Answer (d) is correct. From (7.26) we have: $R_1 \simeq 0$, $R_e = 0$, $S_{T\max} = \dfrac{R_2}{X_T}$

From (7.9) and (7.7) we have:

$$P_m = qR_2\left(\frac{1-s}{s}\right)I_2^2$$

From (7.22):

$$P_m = qR_2\left(\frac{1-s}{s}\right)\frac{V_e^2}{\left(\frac{R_2}{s}\right)^2 + (X_T)^2} = q\left(\frac{R_2}{s} - R_2\right)\frac{V_e^2}{\left(\frac{R_2}{s}\right)^2 + (X_T)^2}$$

$$a \triangleq \frac{R_2}{s} \Rightarrow P_m = q(a - R_2)\frac{V_e^2}{a^2 + X_T^2} \Rightarrow \frac{\partial P_m}{\partial a} = 0 \Rightarrow (a^2 + X_T^2) - 2a(a - R_2) = 0 \Rightarrow$$

$$\Rightarrow a^2 - 2aR_2 - X_T^2 = 0 \Rightarrow a = R_2 + \sqrt{R_2^2 + X_T^2} \Rightarrow \frac{R_2}{s_{P\max}} = R_2 + \sqrt{R_2^2 + X_T^2}$$

$$\Rightarrow s_{P\max} = \frac{1}{1 + \sqrt{1 + \left(\frac{X_T}{R_2}\right)^2}} \Rightarrow s_{P\max} = \frac{1}{1 + \sqrt{1 + \left(\frac{1}{s_{T\max}}\right)^2}}$$

7.3 Asynchronous Machines Analysis in DIgSILENT PowerFactory

This section covers electric machine analysis using engineering software, specifically DIgSILENT PowerFactory, rather than calculators and static methods. This section demonstrates how DIgSILENT PowerFactory, a powerful engineering software, simplifies electrical engineering analysis.

7.3.1 Starting, Disconnecting, Connecting, and Short Circuit

Let´s again consider the IEEE 9 bus network. Here, an asynchronous motor (ASM) with the specifications of Figure 7.12 and like Figure 7.13 is supposed to be added to the network.

It can be seen in Figure 7.13 that an ASM is connected to bus 10 of the network. Since the ASM consumes a significant amount of reactor power, a capacitor can be used in the same bus or nearby buses to compensate for the required reactor power. Now we consider a scenario for this ASM according to Table 7.3. In this scenario, ASM start and stop, short circuit, and short circuit removal are considered. The work steps are as follows in the order of Table 7.3.

1. Initially, the network remains in the steady state for one second.
2. At one second, we energize the asynchronous motor (ASM) by closing its switch.

FIGURE 7.12
Specifications of a 90 MW ASM.

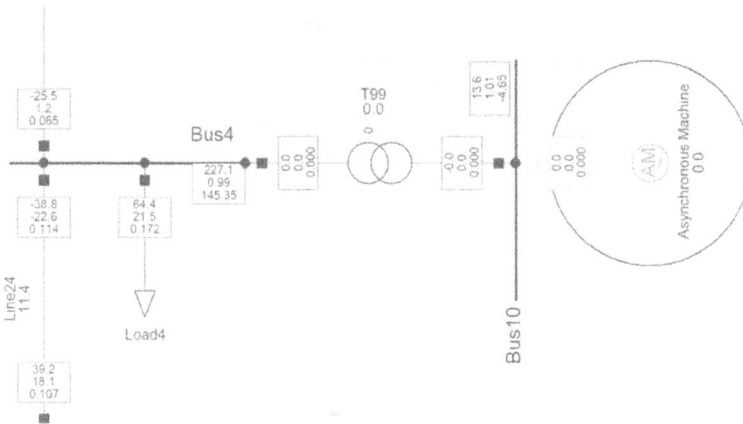

FIGURE 7.13
How to connect the ASM to the network (initially, its circuit is open).

TABLE 7.3

ASM Analysis Scenario

No.	Event	Action	Object	Execution time (Start/Stop)
1	Steady-state	Start Simulation	–	0–1
2	Switch Event (1)	Close	Breaker of ASM	1
3	Switch Event (2)	Open	Breaker of ASM	2
4	Switch Event (3)	Close	Breaker of ASM	5
5	Short-Circuit Event (1)	Three-Phase Short-Circuit	Bus of ASM	8
6	Short-Circuit Event (2)	Clear Short-Circuit	Bus of ASM	8.3
7	Nothing	Continue Simulation	–	8.3–12
8	Stop	Stop Simulation	–	12–12

3. At two seconds, before the motor reaches its nominal speed, we disconnect it by opening the switch.

4. At five seconds, while the motor is still in motion, we reconnect it to the network.

5. At eight seconds, once the motor has reached nominal speed, a three-phase short circuit occurs on the ASM bus.

6. At 8.3 seconds, we clear the short circuit while the motor is still rotating.

7. From 8.3 to 12 seconds, the system continues operation with no further events.

8. At 12 seconds, the simulation concludes.

Now we will examine the outputs of this scenario with emphasis on the parameters of the asynchronous motor and network synchronous generators. Figures 7.14–7.19 show the important network parameters in this scenario. For a better analysis, the event times in the network are presented in the Figures. Figure 7.14 shows the angles of two network generators in degrees relative to the reference generator.

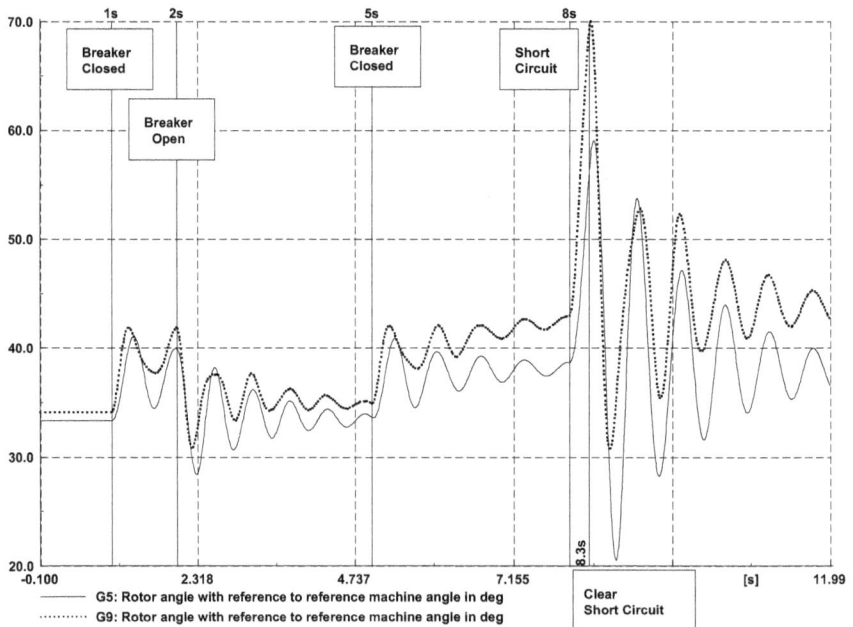

FIGURE 7.14
The angles of two network generators in the scenario.

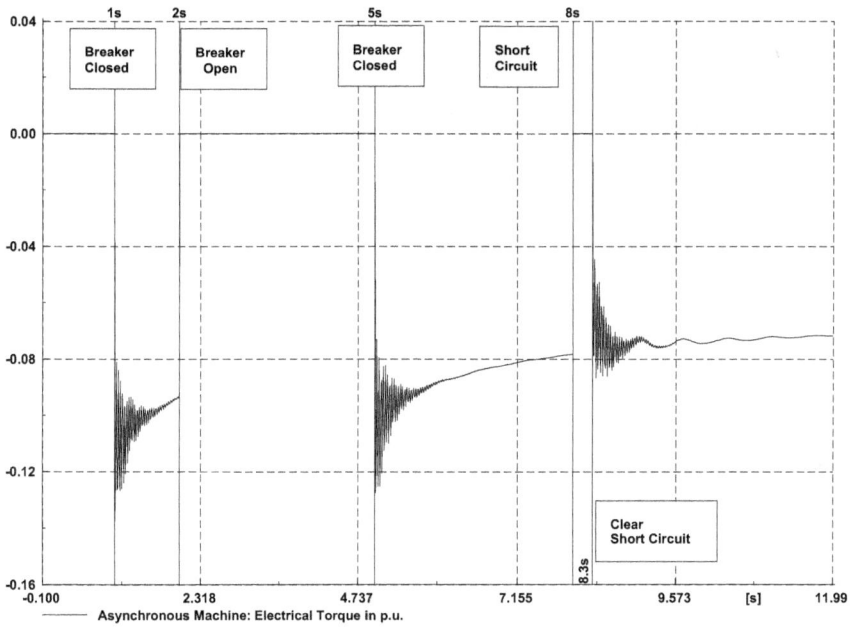

FIGURE 7.15
ASM electric torque in the scenario.

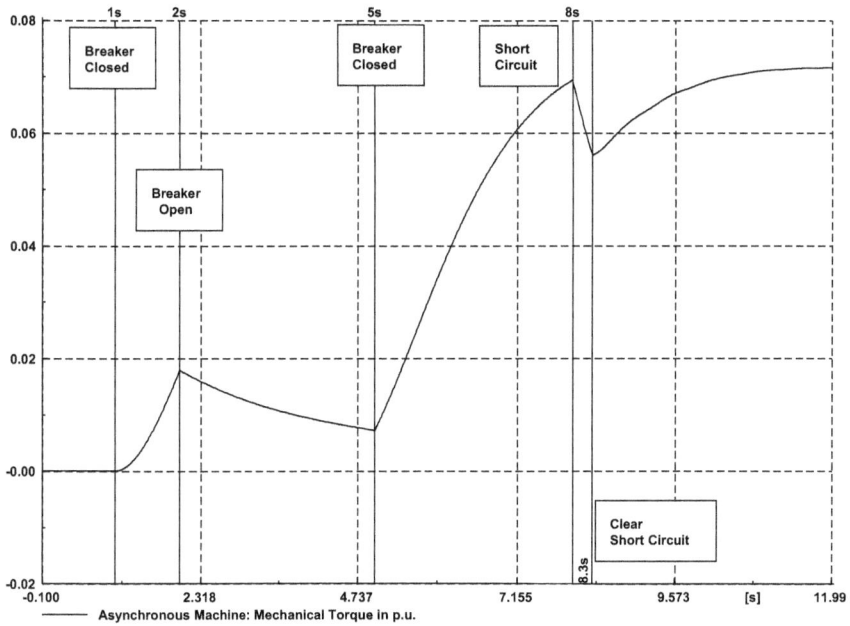

FIGURE 7.16
ASM mechanical torque in the scenario.

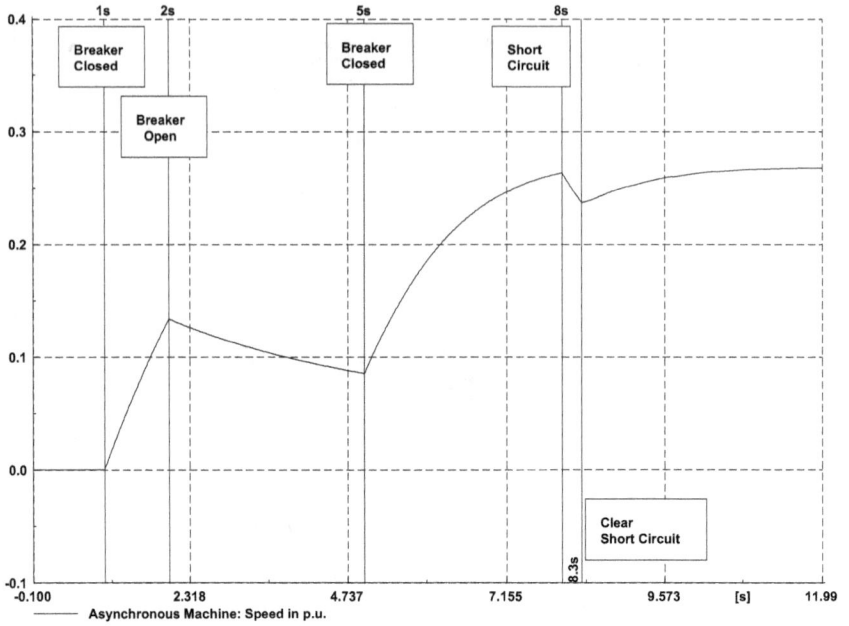

FIGURE 7.17
ASM speed in the scenario.

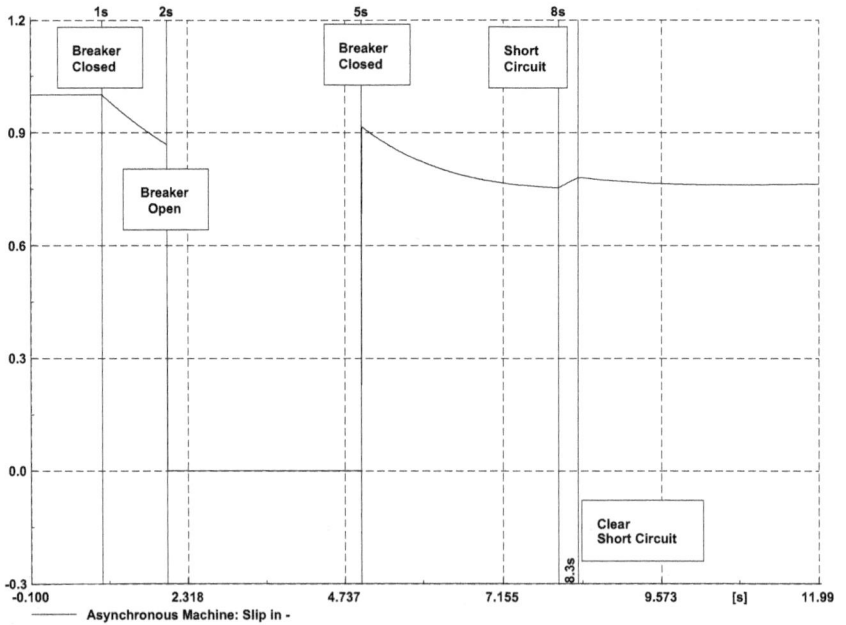

FIGURE 7.18
ASM slip in the scenario.

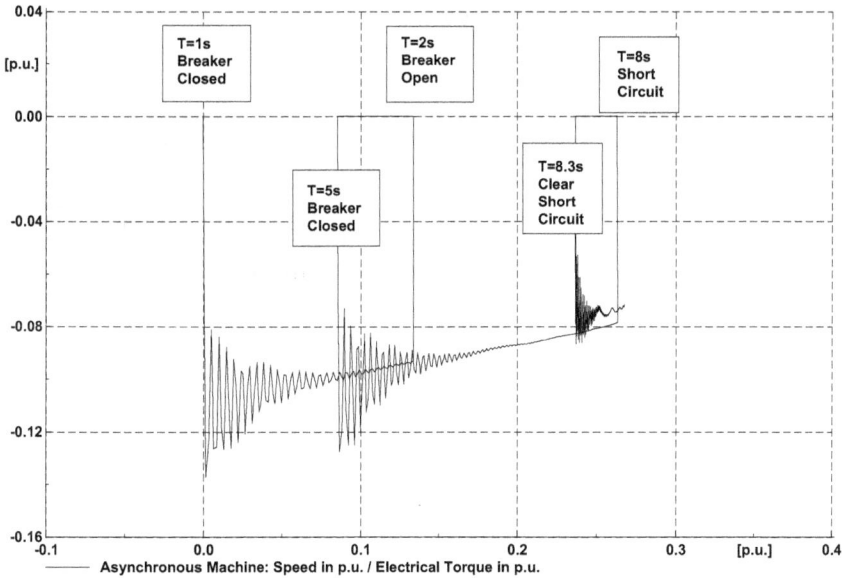

FIGURE 7.19
ASM torque-speed curve in the scenario.

Now we analyze the scenario steps for generator angles according to Table 7.4 and from Figure 7.14.

As can be seen from Table 7.4 and Figure 7.14, there is an obvious difference between the amount of oscillation at the entry and exit of the ASM

TABLE 7.4

Angles of Generators According to the Scenario

Step	Execution Time	Event	Rotor Angles
1	0–1 second	Nothing	The angles of the generators are constant.
2	1 second	Breaker of ASM closed	The angle of the generators increases with positive acceleration and then oscillates.
3	2 seconds	Breaker of ASM opens	The angle of generators decreases with negative acceleration and then oscillates
4	5 seconds	Breaker of ASM closed	The angle of the generators increases with positive acceleration and then oscillates.
5	8 seconds	Short circuit in ASM bus	The angle of generators increases strongly with positive acceleration and then oscillates
6	8.3 seconds	Clear short circuit in ASM bus	The angle of the generators decreases sharply with negative acceleration and then oscillates.
7	8.3–12 seconds	Nothing	The angle of the generators is stable with high oscillation.

with short circuit mode. In fact, like a cardiologist, by examining the electrocardiogram (ECG) of this patient, it is possible to diagnose and predict what happened to the network.

Figures 7.15–7.18 respectively show the electrical torque of the ASM in terms of per unit, mechanical torque, slip, and speed of the ASM in terms of per unit. In all these Figures, the time of occurrence is added.

Our emphasis from examining these Figures is the ability to detect events in the network from the system outputs like a cardiologist. The review summary of Figures 7.15–7.18 is shown in Table 7.5.

As can be seen in Figures 7.15–7.18 and Table 7.5, it is not possible to trace the fault created from the electric torque curve and its oscillations are very high, however, events on the network can be analyzed from the speed, slip, and mechanical torque curves.

In the end, you can see the famous figure of ASM, i.e., torque-speed curve in Figure 7.19. All the events of the scenario can be seen in this figure. All the events have left a good mark. If the events from second 1 to second 8 are followed, the starting of the ASM, its removal and re-entering of the circuit of the ASM, the short circuit, and then the elimination of the short circuit can be easily seen.

It should be noted that in all the previous curves where the parameters were in terms of time, the path of the curve is from left to right, but in Figure 7.19 where the time has been removed in cases where the motor is out of circuit (between second 2 and second 5) curve turns to the left. In addition, when the motor is short-circuited, the curve turns from right to left.

We hope that as we enjoyed analyzing the ECG of our network like a skilled cardiologist, you can continue to do so yourself with the explanations in the next section.

7.4 Appendix, Modeling Steps of Asynchronous Machines in DIgSILENT PowerFactory

The impacts of motor starting, sudden motor start and stop, and short circuits on the motor's bus have been examined in this section by learning how to add an asynchronous motor to the network. In this modeling, the impact of PSS is also examined. This section requires a basic understanding of DIgSILENT PowerFactory software.

Use the file (Chapter8.pfd) to begin this section from the beginning, and use the file (Chapter7.pfd) to view the outcome. Proceed with the remainder of this step after importing the network. It is recommended to do the following 18 steps in order.

TABLE 7.5

ASM Parameters According to the Scenario

Step	Execution time	Event	Electrical Torque (pu)	Mechanical Torque (pu)	Speed (pu)	Slip (pu)
1	0–1 second	Nothing	Constant (0)	Constant (0)	Constant (0)	Constant (1)
2	1 second	Breaker of ASM closed	Starting with a strong swing and slowly increasing	Exponential increase	Increases Linearly	Exponential reduction
3	2 seconds	Breaker of ASM open	Constant (0)	Decreases Linearly	Decreases Linearly	Constant (0)
4	5 seconds	Breaker of ASM closed	Starting with a strong swing and slowly increasing	Increase with negative acceleration	Increase with negative acceleration	At first, the slip increases and then decreases exponentially.
5	8 seconds	Short circuit in ASM bus	Constant (0)	Strong linear reduction	Strong linear reduction	Slip increases exponentially.
6	8.3 seconds	Clear short circuit in ASM bus	Starting with a strong swing and slowly increasing	Increase with low negative acceleration	Increase with low negative acceleration	Slip is slowly reduced
7	8.3–12 seconds	Nothing				

Step 1:

Because asynchronous motors need a significant amount of reactive power, the circuit needs to have a capacitor (No. 1 Figure 7.20).

Step 2:

Disable "**Composite Model**" in generator 5 (out of service). The working method is shown in Figures 7.21 and 7.22.

Step 3:

By pressing the "**Ctrl**" key and selecting bus 9 and transformer T98 (No. 1 and 2 in Figure 7.23), make a copy of it. Then right-click near bus 4 (No. 3) and select the paste option (No. 4). You will get to Figure 7.24. You can move the name of bus 4 for beauty (No. 5 in Figure 7.24).

Note: If the transformer is not connected to bus 4, right-click on it and select the connect option. In the v2021, click Past (New) in No. 4 of Figure 7.23.

Step 4:

Select an asynchronous machine from the right section (left in v2021) (No. 1 Figure 7.25) and connect it to Bus9(1). By right-clicking on an empty space, the asynchronous element is disconnected from the mouse (No. 3 in Figure 7.25). Next, disconnect the asynchronous motor connection key (No. 4 in Figure 7.25). The key becomes hollow. Because the start of the asynchronous motor must be checked, the motor must be disconnected in the initial state and connected during the simulation to see the effect of adding the motor.

Step 5:

Double-click on the asynchronous machine and enter the type information according to the software edition. See Figure 7.26

According to the software edition, as shown in Figures 7.27–7.29, copy a specific type of asynchronous machine from the main library and paste it into your case study library. Don't forget the OK key at the end.

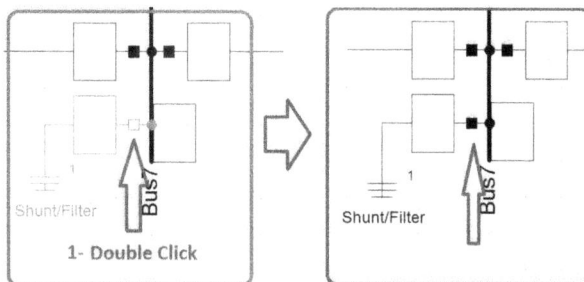

FIGURE 7.20
Reconnecting the capacitor to the grid.

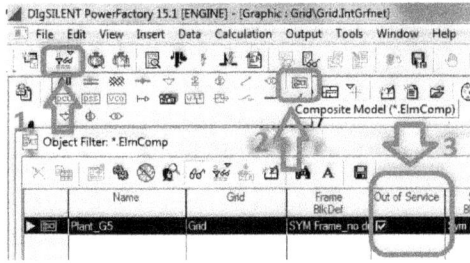

FIGURE 7.21
Disable all controllers (v15.1).

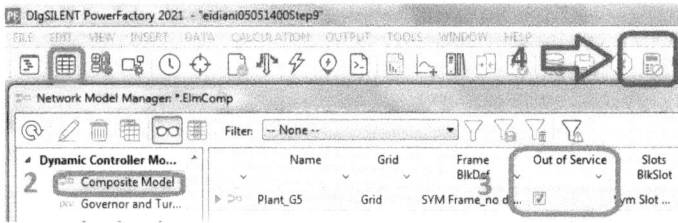

FIGURE 7.22
Disable all controllers (v2021)—if No. 3 is not visible, press the reset key (No. 4).

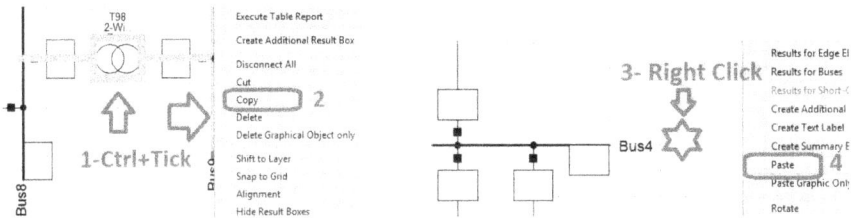

FIGURE 7.23
Copy-paste a bus and transformer.

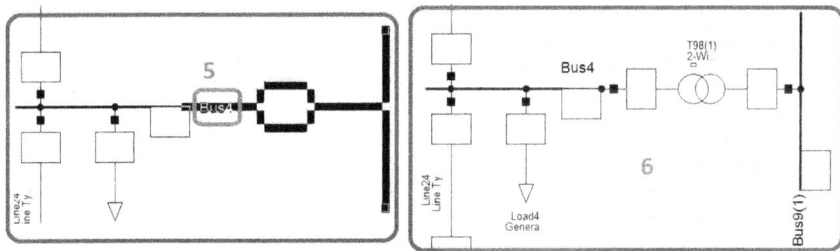

FIGURE 7.24
Bus name transfer.

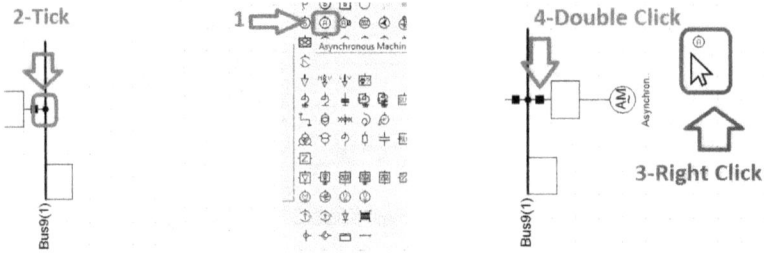

FIGURE 7.25
Connecting the asynchronous motor to the bus and disconnecting it from the network.

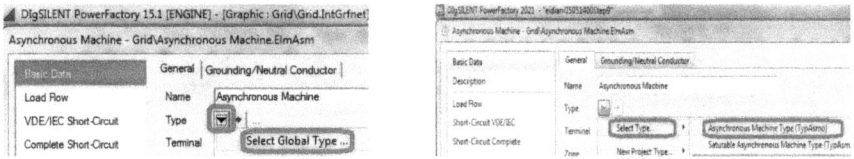

FIGURE 7.26
Enter the type information.

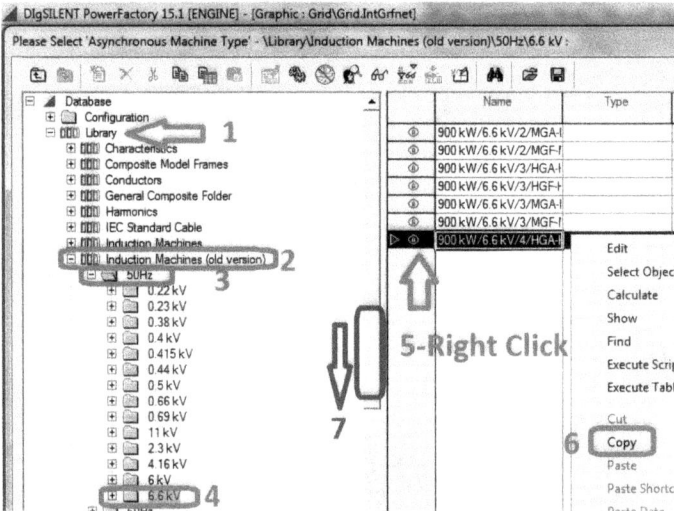

FIGURE 7.27
Selecting an asynchronous machine model from the main library (v15.1).

Step 6:

According to the software edition and Figures 7.30 and 7.31, double-click on the asynchronous machine, press the arrow to the right of the type and change the main information of the machine. In the **Load Flow** section, select Single Cage. Click OK and enter the main page of the software.

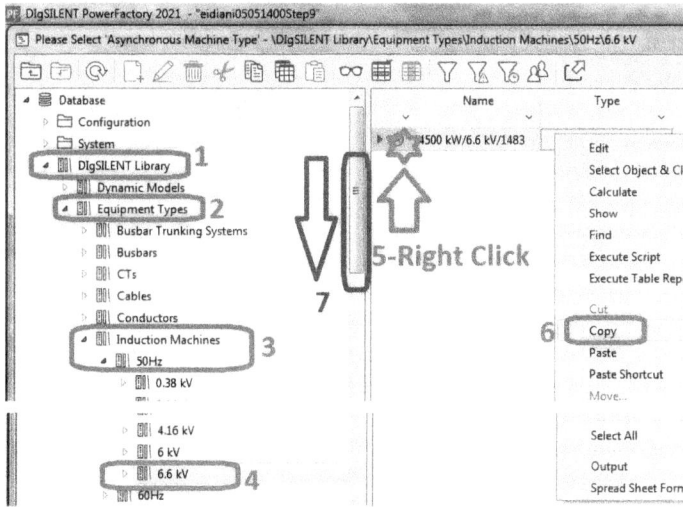

FIGURE 7.28
Selecting an asynchronous machine model from the main library (v2021).

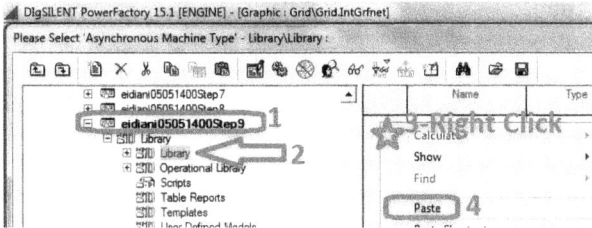

FIGURE 7.29
Paste the selected asynchronous machine model from the main library to the case study library (v15.1, v2021).

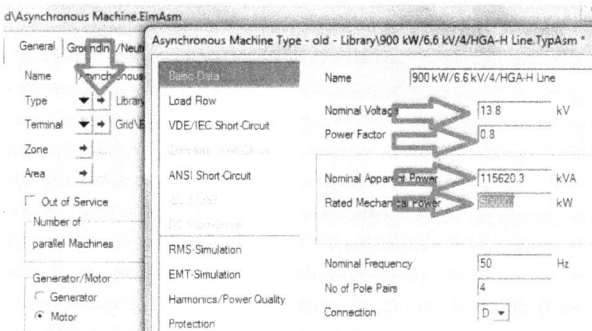

FIGURE 7.30
Enter the information of the type of asynchronous motor (v15.1).

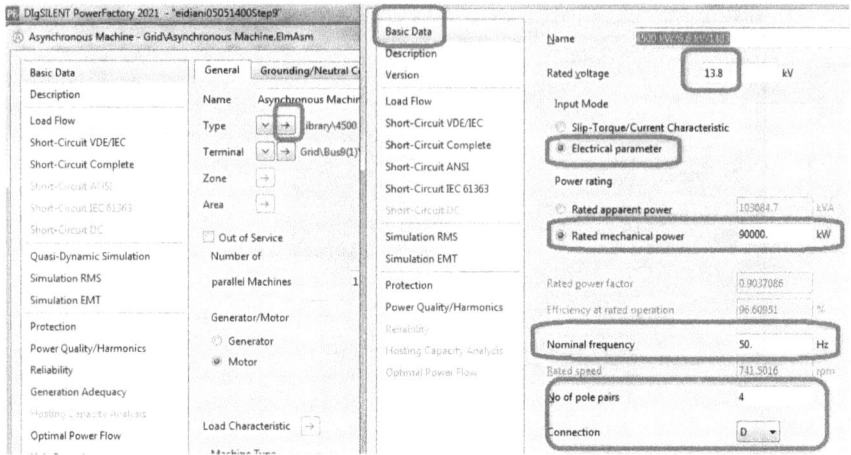

FIGURE 7.31
Enter the information of the type of asynchronous motor (v2021).

Step 7:
Tick the icon (⚡) or (⚡) and clear the short circuit and discon-nection times or tick the "out of service".

Step 8:
Right-click on the asynchronous machine, select **Define**, then **Switch event** and **Basic data**. In the opened window, set the time to one second and click **Close** in Breaker. This action causes the asynchronous motor to start in 1 second. Nothing seems to happen when you click the **Ok** button. Only this information is entered in the (⚡) or (⚡) section.

Step 9:
Repeat step 8 again. In the second 2, check **open** in the Breaker key. This will stop the asynchronous motor in 2 seconds.

Step 10:
Repeat step 8 once more and in second 5, check **close** in Breaker. This action causes the asynchronous motor to re-enter the circuit at 5 seconds while the motor is still rotating.

Step 11:
Right-click on Bus9(1), select **Define** and then **Short-Circuit Event**. In the opened window, set the time to **8** seconds. The pro-gram's **Fault Type** defaults to a three-phase short circuit. If not, select three-phase short circuit. This action causes a short circuit on the terminal of the asynchronous machine in 8 seconds.

Step 12:
Repeat step 11. Set the time to 8.3 seconds and select the **Fault Type** as **Clear Short Circuit**. This action will clear the three-phase short circuit of step 11 on the bus.

Step 13:

If the above steps are done correctly, you should have information like Figure 7.32 in ⚡ or ⚡ section or **Edit Simulation Events**.

Step 14:

The variables of the asynchronous machine that you wish to view in the simulation should be visible in this area. **Right-click** on the asynchronous machine and select **Define**. Next, select **Results** for **RMS/EMT Simulation**. In Figure 7.33, double-click the asynchronous machine.

Refer to Figures 7.34 and 7.35 for versions 15.1 and 2021, respectively. Choose the variables **slip** (No. 3), mechanical torque (**xmt**), and electric torque (**xme**) in the **EMT-Simulation** or **RMS-Simulation** window (No. 1). These variables will be moved to the **Selected Variable** section (No. 4) when you check (No. 3). Choose the **Signal** option in the **Variable Set** section (No. 2) of the same page, then add the **speed pu** variable to the variables that have been chosen (No. 4).

Step 15:

To create a plot page for v(15.1), adhere to the instructions shown in Figure 7.36. Choose the **Append new VI(s)** icon (🔁) on the page that opens, then choose the **Subplot (VisPlot)** option in 4 plots and click the "**OK**" button. Four square graphs can be seen by using the (🔲) key.

Follow the steps mentioned in Figures 7.37 and 7.38 to make a graphic page for v(2021).

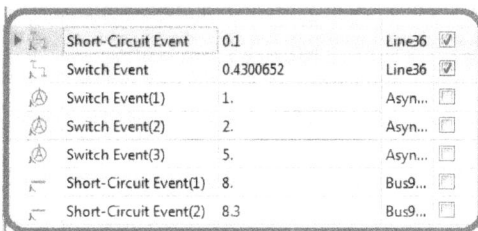

FIGURE 7.32
Edit simulation.

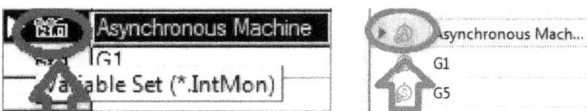

FIGURE 7.33
Variable set in asynchronous machine.

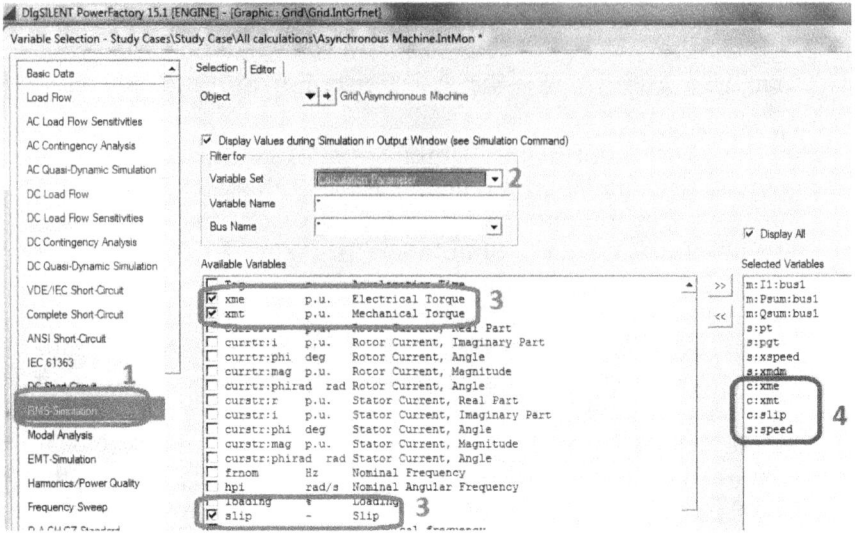

FIGURE 7.34
Define asynchronous motor variables to display in the simulation (v15.1).

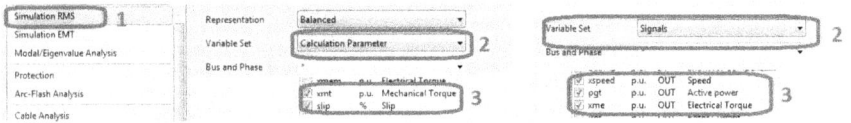

FIGURE 7.35
Define asynchronous motor variables to display in the simulation (v2021).

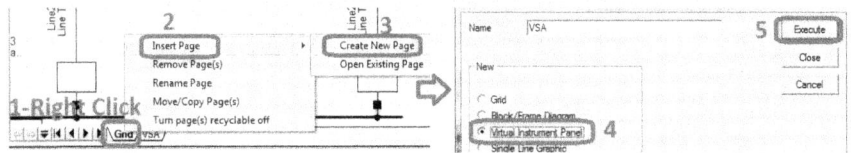

FIGURE 7.36
Creating a plot page or Virtual Instrument Panel (v15.1).

Click **OK** after choosing the **Insert Plot** icon (No. 4 in Figure 7.38), then the **Curve Plot** option (numbers 5 and 6 in Figure 7.38). Figure 7.38's numbers 4–6 should be repeated four times. Four square graphs can be seen by using the (🖼) key.

Step 16:
In step 15, right-click on each of the four plots and select **Edit** (or double-click). To complete the blanks, double-click on them

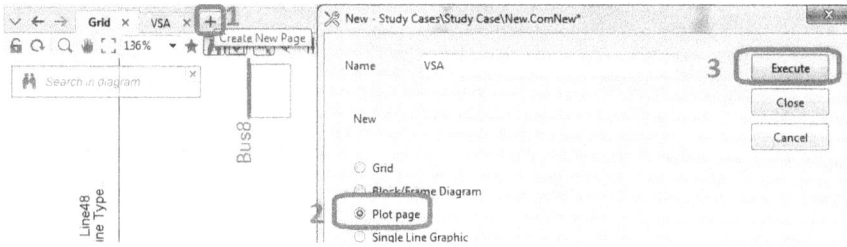

FIGURE 7.37
Creating a plot page (v2021), part 1.

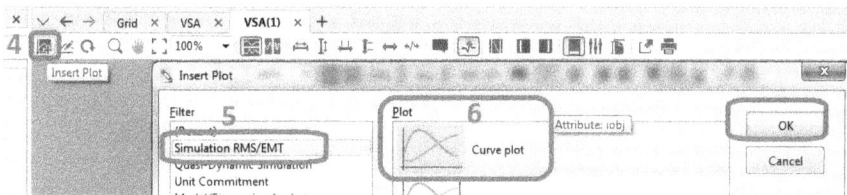

FIGURE 7.38
Creating a plot page (v2021), part 2.

as shown in Figure 7.39. A variable must be represented by each figure. One of the four variables (slip, xme, xmt, and speed) is No. 3 in Figure 7.39.

Step 17:

See Figures 7.40 and 7.41. Once again, follow steps 15 and 16, but this time, utilize an **X-Y plot** rather than a **Subplot**. Select the electrical torque (xme) variable in the *y*-axis section and the speed variable in the *x*-axis section. This figure displays the well-known torque-speed curve.

Step 18:

We are prepared to simulate at this point. Locate the toolbox section icons in the center of the software in Figure 7.42. We ensure that the constructed system can determine the initial working point by selecting the **Calculate Initial Condition** button (No. 1).

FIGURE 7.39
Defining variables on the graphics page (v15.1 and v2021).

Variables:

	Result File ElmRes	Element y-Axis	Var. y-Axis	Var. x-Axis
▶ 1	All calculations	Asynchronous Machine	c:xme	s:speed

FIGURE 7.40
Definition of *X-Y* variables (v15.1).

Curves:

	Visible	Element Y-Axis	Variable Y-Axis	Element X-Axis	Variable X-Axis
1	✓	Asynchronous M...	s:xme	Asynchronous M...	s:xspeed

FIGURE 7.41
Definition of *X-Y* variables (v2021).

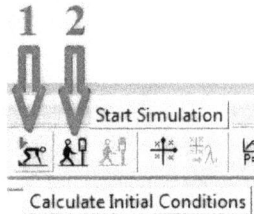

FIGURE 7.42
Calculate initial condition and start simulation keys.

Figures 7.43–7.45 show the results of setting the simulation time to 12 seconds and pressing the **Start Simulation** key (No. 2).

Examine the output plots and look for more asynchronous machine settings.

7.5 Summary

The three-phase symmetrical asynchronous motor, known for its cost-effectiveness and great longevity, is the most prevalent motor type due to its efficiency and stable performance. It operates on a balanced three-phase power supply, producing a uniform magnetic field that leads to effective rotor current distribution. These motors are widely utilized in various industrial applications, including pumps and generators, and can be converted into induction generators by exceeding synchronous speed. The chapter outlines the classical analysis of these motors and generators, including simulations using DIgSILENT PowerFactory software, providing insights into their operation and analysis techniques.

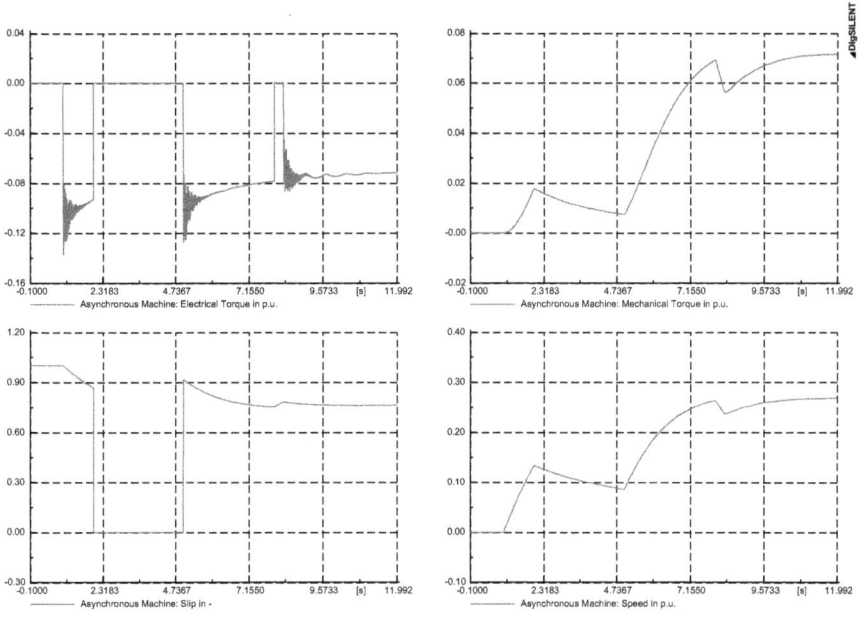

FIGURE 7.43
Figure 4 made in step 16.

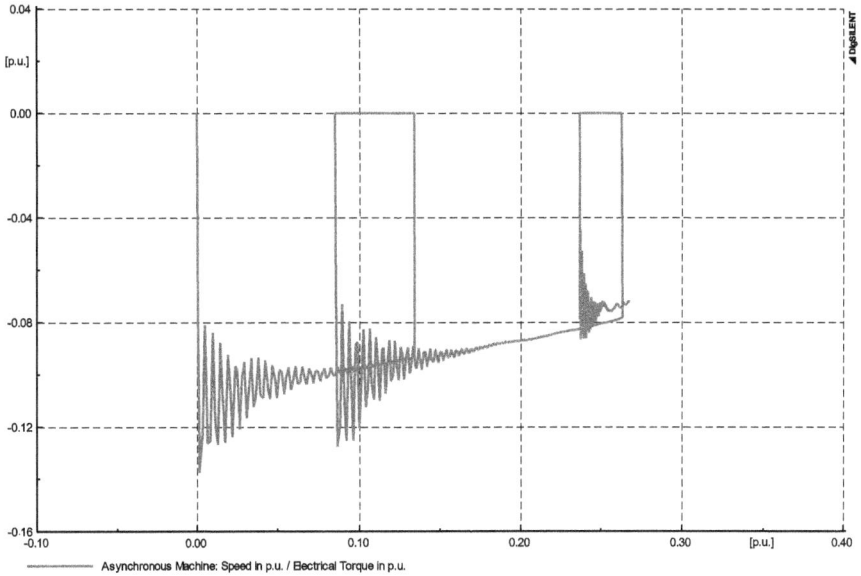

FIGURE 7.44
X-Y torque-speed output figure created in step 17.

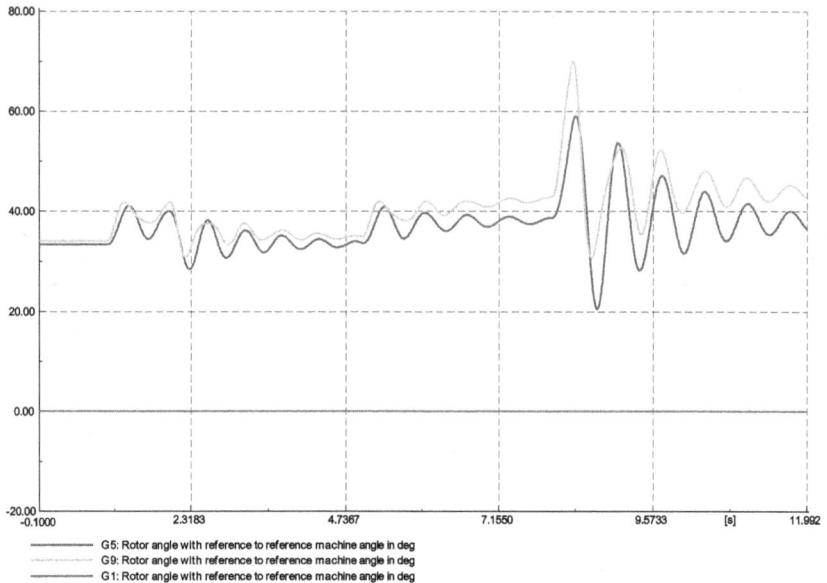

FIGURE 7.45
Changing the angle of the generators after connecting the asynchronous machine.

References

1. Mehrizi-Sani, A., *Electrical Machines and Their Applications*, CRC Press, 2024.
2. Boldea, I., Tutelea, L.N., *Electric Machines*, vol 2, CRC Press, 2021.
3. Patrick, D.R., Fardo, S.W., Richardson, R., Chandra, V., *DC/AC Electrical Fundamentals*, CRC Press, 2024.
4. Gonen, T., *Electrical Machines with MATLAB*, CRC Press, 2012.
5. Gieras, J.F., *Electrical Machines, Fundamentals of Electromechanical Energy Conversion*, CRC Press, 2016.
6. Sen, P.C., *Principles of Electric Machines and Power Electronics*, 3rd edition, Wiley, 2013.
7. El-Hawary, M.E., *Principles of Electric Machines with Power Electronic Applications*, 2nd edition, Wiley IEEE, 2002.
8. Kothari, D.P., Nagrath, I.J. *Electric Machines*, McGrawHill, 2017.
9. Asanbayev, V. *Asynchronous Machines, Basic Calculation Elements from Field Equations*, Springer, 2022.
10. Boldea, I., *Induction Machines Handbook, Steady State Modeling and Performance*, CRC Press, 2020.
11. Razik, H., *Handbook of Asynchronous Machine with Variable Speed*, Wiley, 1988.
12. Beigi, H. M. C., Karami, E., Afjei, E., Rouzbehi, K., "Numerical and experimental investigation of an improved flux path brushless-DC machine for variable speed applications," *IEEE Transactions on Transportation Electrification*, vol. 4, no. 4, pp. 877–887, 2018.

8

Power System Harmonic Analysis

8.1 Introduction

The voltage and current waveforms in power systems typically deviate from the ideal sinusoidal state. Saturation of the transformer core, thyristor control devices, and particularly AC/DC converters can all be the cause of this disruption or departure from the pure sine wave. With the aid of Fourier series, we can break down these non-sine waves into their constituent parts, which we refer to as harmonic. Harmonics are multiples of the synchronous frequency's main wave. Although non-integer harmonics, also known as hidden harmonics, can exist under specific network conditions, the power system typically only displays harmonics that are an integer multiple of the fundamental frequency. "Even harmonics" are absent from a symmetrical network. As a result, compared to odd harmonics, even harmonics are either extremely rare or unlikely to be observed in power systems.

The harmonic components of all three phases in symmetrical power systems exhibit a phase shift (φ_f =120 ($f_h \div f_1$)). Therefore, by connecting the transformer's triangle, harmonics that are multiples of three, such as zero components, can be stopped because they are in phase with one another. The aforementioned premises state that the odd and non-multiple harmonic orders in symmetric power systems are typically three, $i.e.$, 5, 7, 11, 13, 17, 19, 23, 25, 29, 31, 35, 37, etc.

Harmonic concepts are briefly covered in this chapter. Next, DIgSILENT PowerFactory software is used to analyze a small harmonic network. How to use the software and the possibility for further work are explained in the appendix. For more detailed work and further work, you can refer to the references [1–9] at the end of this chapter.

8.2 Harmonic Indices

Electric power quality is determined by a variety of indices. Below is a list of some of the indices that DIgSILENT software uses. Although they are defined for electric current, these indicators also apply to voltage.

DOI: 10.1201/9781003590514-8

The principal component at synchronous frequency, first, fundamental, or nominal harmonic, is all specified as follows.

$$I_{f1} \tag{8.1}$$

Current in the ith harmonic: I_{fi} $\hfill$ (8.2)

$$\text{RMS current: } I_{\text{rms}} = \sqrt{\sum_{i=1}^{i=\infty} I_{fi}^2} \tag{8.3}$$

Harmonic distortion in the ith harmonic: $\text{HD}_i = \dfrac{I_{fi}}{I_{f1}}$, $\;1:\text{Nominal}$ $\quad$ (8.4)

$$\text{Total harmonic distortion: } \text{THD}\% = \frac{100}{I_{f1}} \sqrt{\sum_{i=2}^{i=\infty} I_{fi}^2} \tag{8.5}$$

Distortion index in the ith harmonic: $\text{DIN}_i = \dfrac{I_{fi}}{I_{\text{rms}}}$ $\quad$ (8.6)

Total distortion index (TDIN) or Total demand distortion (TDD)

$$\text{TDD}\% \triangleq \text{TDIN}\% = \frac{100}{I_{\text{rms}}} \sqrt{\sum_{i=2}^{i=\infty} I_{fi}^2} \tag{8.7}$$

Active (real) power in the ith harmonic: $P_{fi} = R_e\left((\hat{V}_{fi})(\hat{I}_{fi}^*)\right)$ $\quad$ (8.8)

Reactive power in the ith harmonic: $Q_{fi} = I_m\left((\hat{V}_{fi})(\hat{I}_{fi}^*)\right)$ $\quad$ (8.9)

Apparent power in the ith harmonic: $S_{fi} = \sqrt{P_{fi}^2 + Q_{fi}^2}$ $\quad$ (8.10)

Power factor in the ith harmonic: $\cos\left(\varphi_{fi}\right) = \dfrac{P_{fi}}{Q_{fi}}$ $\quad$ (8.11)

$$\text{Total active power: } \text{TP} = \sum_{i=1}^{i=\infty} P_{fi} \tag{8.12}$$

Total apparent power: $\text{TS} = (V_{\text{rms}})(I_{\text{rms}})$ $\quad$ (8.13)

Total reactive power: $\text{TQ} = \sqrt{\text{TS}^2 - \text{TP}^2}$ $\quad$ (8.14)

$$\text{Total power factor: } T\cos\varphi = \frac{\text{TP}}{\text{TS}} \qquad (8.15)$$

Creating a harmonic generation source in a harmonic load is the most straightforward method of producing harmonics in a network. Harmonics and symmetrical load are used for simplicity. The ability to compute the aforementioned harmonic relations using a calculator is essential for a deeper comprehension of harmonics, as demonstrated in the example that follows.

Example 8.1

Table 8.1 provides information about a harmonic bus. Non-harmonic information is shown in the first column, first or fundamental harmonic information is shown in the second, fifth harmonic information is shown in the third, and seventh harmonic information is shown in the fourth. The seventh harmonic of the current is 14.285% of the main current, while the fifth harmonic is 20%. Below, we'll look at the harmonic index relationships.

From (8.3) for current:

$$I_{rms} = \sqrt{\sum_{i=1}^{i=\infty} I_{fi}^2} = \sqrt{(244.43995)^2 + (48.88799)^2 + (34.91825)^2} = 0.25171\,\text{kA}$$

From (8.3) for voltage:

$$V_{rms} = \sqrt{\sum_{i=1}^{i=\infty} V_{fi}^2} = \sqrt{(224.07241)^2 + (67.16586)^2 + (31.90625)^2} = 236.0883\,\text{kV}$$

From (8.4) for voltage:

$$HD_1 = 100\frac{224.07241}{224.07241} = 100\%, \; HD_5 = 100\frac{67.16586}{224.07241} = 29.975\%,$$

$$HD_7 = 100\frac{31.90625}{224.07241} = 14.239\%$$

From (8.4) for current:

$$HD_1 = 100\frac{244.43995}{244.43995} = 100\%, \; HD_5 = 100\frac{48.88799}{244.43995} = 20\%,$$

$$HD_7 = 100\frac{34.91825}{244.43995} = 14.285\%$$

From (8.5) for current:

$$THD_I = \frac{100}{I_{f1}}\sqrt{\sum_{i=2}^{i=\infty} I_{fi}^2} = \frac{100}{244.43995}\sqrt{(48.88799)^2 + (34.91825)^2} = 24.5777\%$$

TABLE 8.1

Information About a Harmonic Busbar

Column 1	Column 2	Column 3	Column 4
Without Harmonic	Harmonic 1 (Fundamental)	Harmonic 5	Harmonic 7

Column 1 — Without Harmonic

Bus 4

Ui=224.072
u=0.97423
phiu=145.154

P=90.000
Q=30.000
I=0.24444
phii=−126.71919

Harmonics

	I_hf_I %	phi_h+i*phi_i deg
Bar5	20	0.
Bar7	14.285	0.

Load 4

Column 2 — Harmonic 1 (Fundamental)

Bus4

THD=33.185 %
HD=100.000 %
UIrms=236.08837 kV
UI=224.07241 kV

I=244.43995 A
Irms=0.25171 kA
TP=89.82054 MW
TQ=50.26858 Mvar
TS=102.93036 MVA
Tcosphi=0.87263
THD=24.57766 %
HD=100.00000 %
P=89999.98750 kW
Q=29999.99443 kvar
S=94868.31619 kVA

Load4

Column 3 — Harmonic 5

Bus4

THD=33.185 %
HD=29.975 %
UIrms=236.08837 kV
UI=67.16586 kV

I=48.88799 A
Irms=0.25171 kA
TP=89.82054 MW
TQ=50.26858 Mvar
TS=102.93036 MVA
Tcosphi=0.87263
THD=24.57766 %
HD=20.00000 %
P=−121.24047 kW
Q=−5686.07666 kvar
S=5687.36908 kVA

Load4

Column 4 — Harmonic 7

Bus4

THD=33.185 %
HD=14.239 %
UIrms=236.08837 kV
UI=31.90625 kV

I=34.91825 A
Irms=0.25171 kA
TP=89.82054 MW
TQ=50.26858 Mvar
TS=102.93036 MVA
Tcosphi=0.87263
THD=24.57766 %
HD=14.28500 %
P=−58.21688 kW
Q=1928.81737 kvar
S=1929.69575 kVA

Load4

From (8.5) for voltage:

$$\text{THD}_V = \frac{100}{V_{f1}} \sqrt{\sum_{i=2}^{i=\infty} V_{fi}^2} = \frac{100}{224.07241} \sqrt{(67.16586)^2 + (31.90625)^2} = 33.185\%$$

From (8.12):

$$\text{TP} = \sum_{i=1}^{i=\infty} P_{fi} = (89999.98750\,\text{kW}) + (-121.24047\,\text{kW}) + (-58.21688\,\text{kW})$$

$$= 89.82054 \text{ MW}$$

From (8.13) for three-phase:

$$\text{TS} = \sqrt{3}(V_{\text{rms}})(I_{\text{rms}}) = \sqrt{3}(236.08837 \text{ kV})(0.25171 \text{ kA}) = 102.9285 \text{ MVA}$$

$$\cong 102.93036 \text{ MVA}$$

From (8.14):

$$\text{TQ} = \sqrt{\text{TS}^2 - \text{TP}^2} = \sqrt{(102.93036)^2 - (89.82054)^2} = 50.2686 \text{ MVAr}$$

From (8.15)

$$T\cos\varphi = \frac{TP}{TS} = \frac{89.82054}{102.93036} = 0.8726$$

8.3 Harmonic Indices in DIgSILENT

This section demonstrates how to use DIgSILENT PowerFactory software to simulate a basic harmonic network. You can read and follow the appendix section of this chapter to repeat these simulations.

Consider the 9-bus network in Figure 8.1. According to Figure 8.2, load 4 is the only source of harmonic production in this network. It is evident that a symmetrical load with only odd and non-multiple harmonics is assumed. Keep in mind that the harmonics are typically reduced proportionate to their order if we do not have information about the harmonic measurement. One-fifth, or 20%, is the fifth harmonic; one-seventh, or 14.285%, is the seventh; and so forth.

The bus voltage THD of this network is visible in Figure 8.3, and an intriguing conclusion was drawn.

The most significant finding from Figure 8.3 is that, even though the fourth bus is the only one with a harmonic source, the third bus has the highest THD. The short answer is that harmonics moved throughout the network due to the transmission lines' capacitors. Harmonics raise

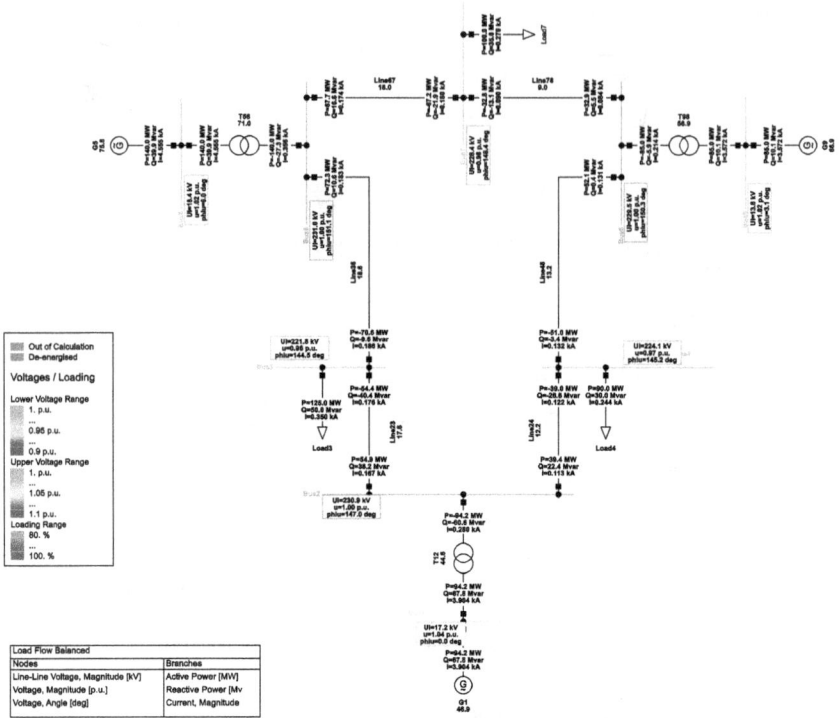

FIGURE 8.1

The load flow output of the 9-bus system (v2021).

Harmonics:

	l_h/l_1 %	phi_h deg
f/fn=5	20.	0.
f/fn=7	14.285	0.
f/fn=11	9.09	0.
f/fn=13	7.692	0.
f/fn=17	5.882	0.
▶f/fn=19	5.263	0.

FIGURE 8.2

Harmonic value table in load 4.

the current in the capacitors by lowering the impedance of the network's parallel capacitors. In actuality, capacitors draw harmonics to themselves and function similarly to a well for harmonics.

HD is displayed for every bus in Figure 8.4. Due to the resonance, the harmonic value of voltages 11 and 13 has increased even though the

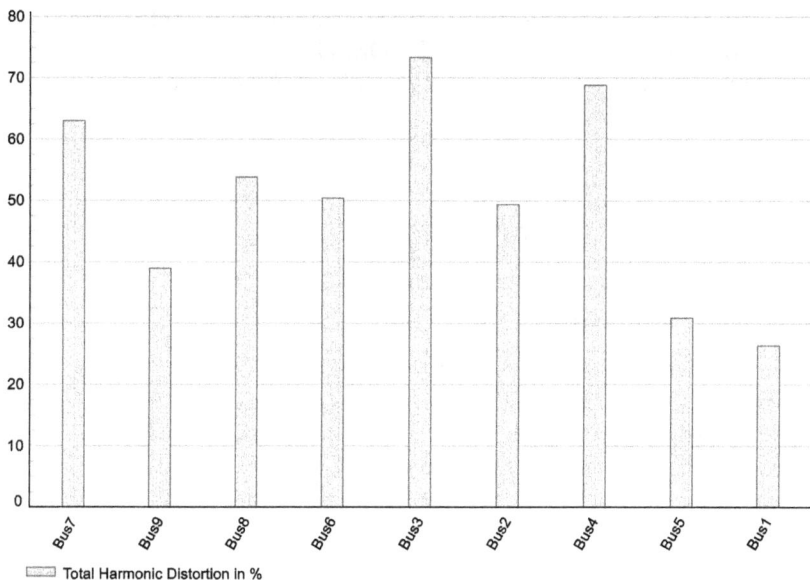

FIGURE 8.3
THD graphic output of the harmonic load flow of the 9-bus network.

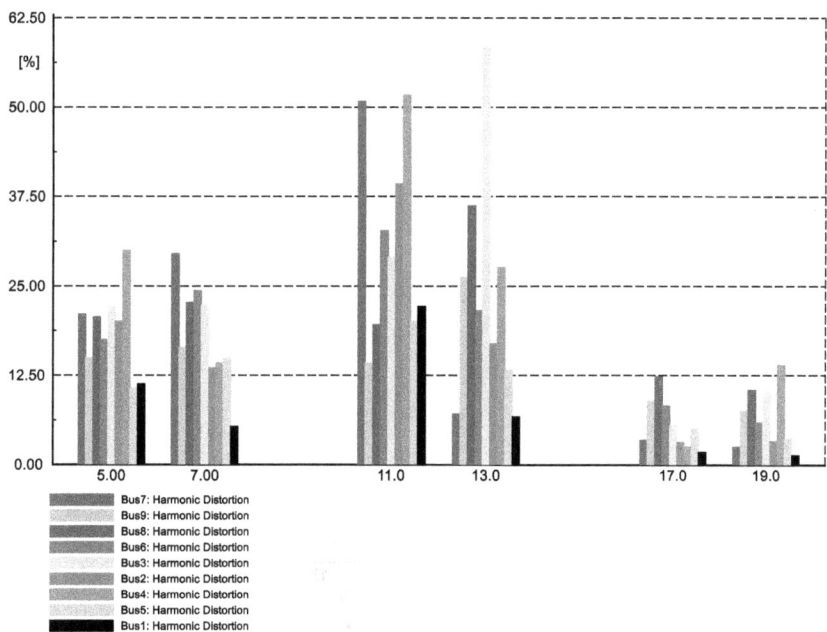

FIGURE 8.4
HD graphic output of the harmonic load flow of the 9-bus network.

harmonic value of currents 11 and 13 injected into the network through bus 4 is lower than harmonics 5 and 7. Drawing the impedance frequency characteristics (frequency sweep) of the following section makes it simple to analyze the resonance effect.

8.4 Impedance Frequency Characteristics (Frequency Sweep)

Frequency sweep is one of DIgSILENT software's most crucial features. Plotting each bus's observed Thévenin impedance against all frequencies is necessary to identify the bus's most significant harmonic. At the frequency where the impedance value is highest, resonance takes place. If necessary, this frequency should be the most crucial one to consider, e.g., moving capacitors, adding a filter, etc.

Figure 8.5 displays the frequency impedance of Bus 4 graphically. It is evident that in four points, specifically in harmonics close to 6, 11, 13, and 19, the impedance has peaked. Resonance has previously been observed in the 11th and 13th harmonics. Additionally, because of this, the 19th

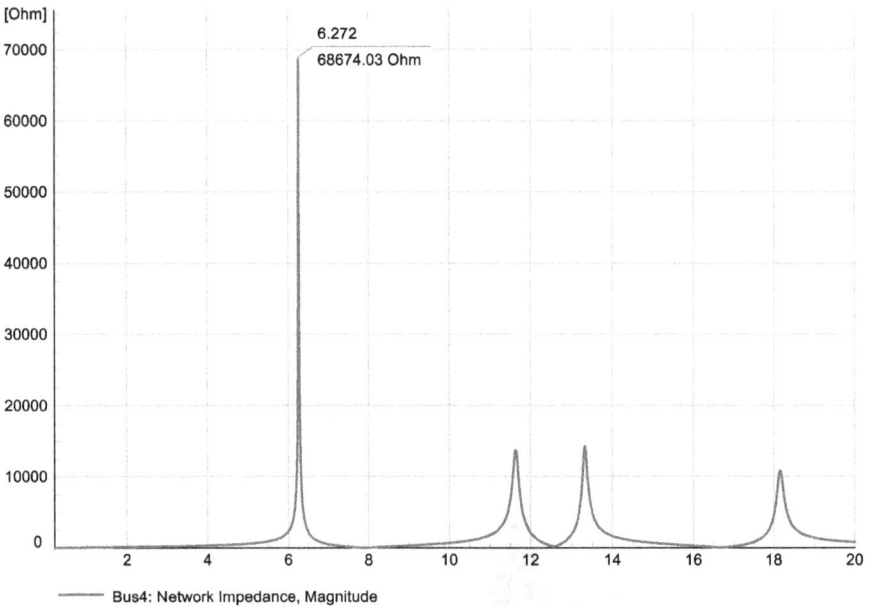

FIGURE 8.5
Graphic output of frequency impedance from bus 4 of 9-bus network.

harmonic is more significant than the 17th. The impact of resonance will be further explored in the following. The impedance of Thevenin reaches its maximum value in Figure 8.4 at the harmonic order value of 6.272. In other words, the network has the highest resonance in this harmonic order. We create a very small number of harmonics in this harmonic order (1%) and examine their impact to observe the resonance effect, even though it is extremely unlikely to be produced. The values of harmonics 5–19 range from 20% to 5%, as shown in Figure 8.2, and the harmonic value of 1% is extremely low for harmonic 6.272.

After adding a very low harmonic value of the resonant current (6.272), the network's HD output is displayed in Figure 8.6. Figures 8.4 and 8.6 can be compared to clearly see the resonance effect. It is evident that the resonant frequency is the most hazardous one, and if the network is secured for this frequency, the network will dampen the remaining harmonics.

This section examines how adding a capacitor to the network affects the resonance frequency. It was mentioned in the earlier sections that the reason for the change in harmonics is capacitors. Bus 7 is therefore equipped with a 20 MVar capacitor. The frequency impedance scan with a capacitor present is displayed in Figure 8.7. There is a noticeable difference in the resonance points when compared to Figure 8.5. This comparison demonstrates the crucial impact that capacitor presence has on harmonics.

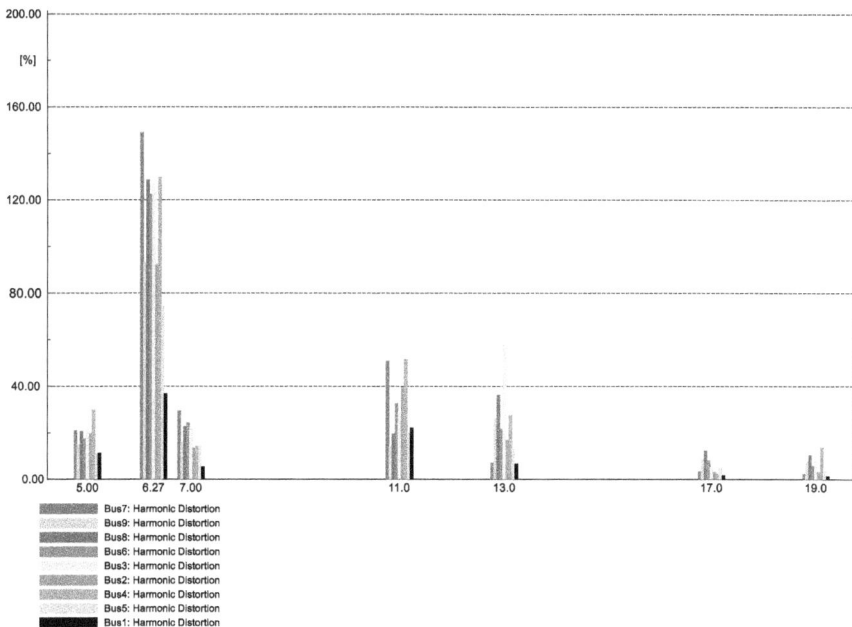

FIGURE 8.6
HD graphic output of the harmonic load flow of the 9-bus network with resonance.

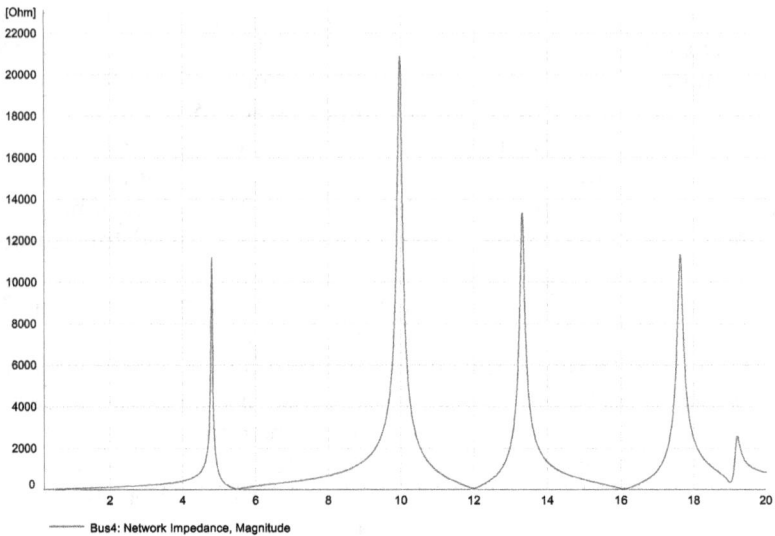

FIGURE 8.7
Graphic output of frequency impedance from bus 4 of 9-bus network with capacitor.

8.5 Two-Choice Questions (Yes/No)

1. Harmonics are sinusoidal waveforms with frequencies that are integer multiples of the fundamental frequency.
2. The fundamental frequency in a power system is typically 50 Hz or 60 Hz.
3. Harmonics cannot cause overheating of transformers and motors.
4. Harmonic distortion can lead to increased power losses in transmission and distribution systems.
5. Power factor correction capacitors can exacerbate harmonic problems.
6. Nonlinear loads, such as rectifiers and inverters, are major sources of harmonics.
7. Fluorescent lighting (with electronic ballasts) is a significant source of harmonic distortion.
8. Induction motors are a major source of harmonic distortion.
9. Power electronic devices, like those used in variable speed drives, can introduce harmonics into the power system.
10. Harmonic distortion cannot be caused by unbalanced loads.

11. Harmonics can cause interference with communication systems.

12. Harmonic distortion can lead to resonance in power systems, causing voltage fluctuations.

13. Harmonics can reduce the efficiency of power systems.

14. Harmonic distortion can cause premature aging of electrical equipment.

15. Harmonics cannot lead to incorrect readings on metering equipment.

16. Harmonic filters are used to attenuate specific harmonic frequencies.

17. Active filters can dynamically compensate for harmonic currents.

18. Passive filters are more expensive than active filters.

19. Proper grounding techniques can help mitigate harmonic problems.

20. Harmonic distortion can be reduced by using power factor correction capacitors.

21. IEEE 519 is a standard that sets limits for harmonic distortion in power systems.

22. IEC 61000-3-2 is an international standard that specifies limits for harmonic currents emitted by equipment connected to the public power supply.

23. Regulatory bodies in many countries have established limits for harmonic distortion in power systems.

24. Compliance with harmonic standards can reduce the risk of power quality problems.

25. Harmonic mitigation measures can be costly and complex to implement.

26. Power quality analyzers cannot be used to measure harmonic distortion.

27. Time-domain analysis can be used to identify harmonic frequencies and magnitudes.

28. Frequency domain analysis is the most common method for analyzing harmonic distortion.

29. Harmonic distortion can be quantified using Total Harmonic Distortion (THD).

30. Continuous monitoring of harmonic levels is essential for effective harmonic management.

8.5.1 Key Answers to Two-Choice Questions

Yes	1, 2, 4–7, 9, 11–14, 16, 17, 19, 21–25, 28–30
No	3, 8, 10, 15, 18, 20, 26, 27

8.6 Appendix, Harmonic Analysis in DIgSILENT

This section requires a basic understanding of DIgSILENT PowerFactory software. You must download the file (Chapter8.pfd) from the book's end-of-book attachments in order to follow this section. Proceed with the remainder of this step after importing the network. It is recommended to do the following steps.

8.6.1 Drawing the THD and HD Diagrams

Step 1:
> Double-click Load4, follow the instructions shown in Figure 8.8 and then press the Ok key once. As a result, load4 becomes a harmonic-producing load.

Step 2:
> Figure 8.9 indicates that you should arrive at or navigate to the first page of load in this section. Complete the table in Figure 8.2 by following the steps in Figure 8.9 and, in the final step, setting *n* to 5.

Step 3:
> In this section, to calculate harmonic load flow, as illustrated in Figure 8.10, navigate to the harmonic section of the Toolbox.

Step 4:
> Execute the harmonic load flow without changing the parameters of Figure 8.11.
> If you have done the steps correctly, you should reach the output of Figure 8.12.

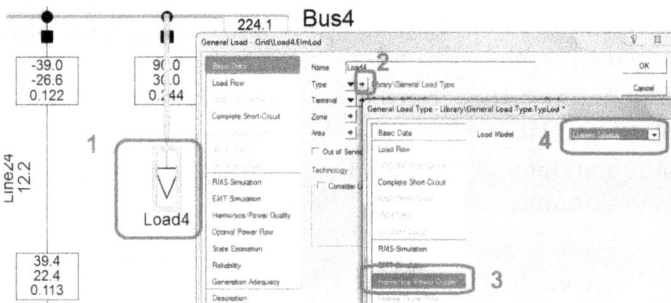

FIGURE 8.8
Converting a load to a harmonic source.

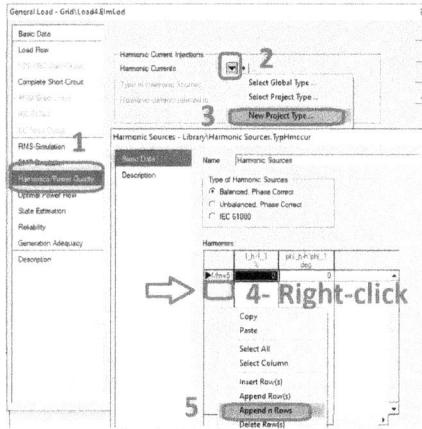

FIGURE 8.9
Entry of harmonic information of Busbar 4.

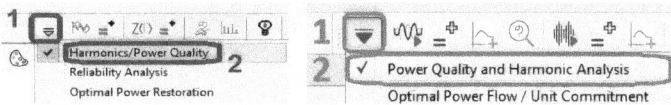

FIGURE 8.10
Access to Harmonic Toolbox in two editions v15.1 and v2021.

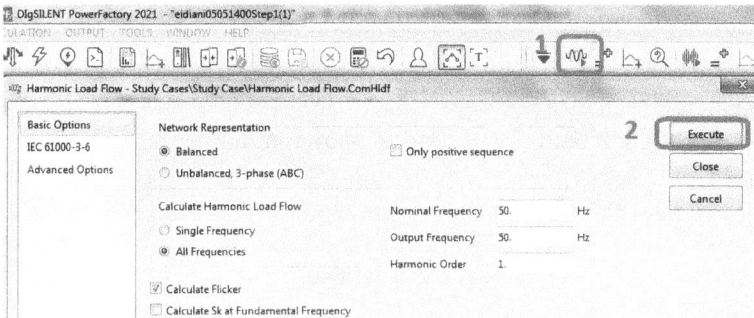

FIGURE 8.11
Harmonic load flow window.

Step 5:

To view the buses' harmonic information, select all buses by using the Ctrl-A key, then right-click on a bus. Figure 8.13 shows that drawing THD or HD buses in a single figure is simple.

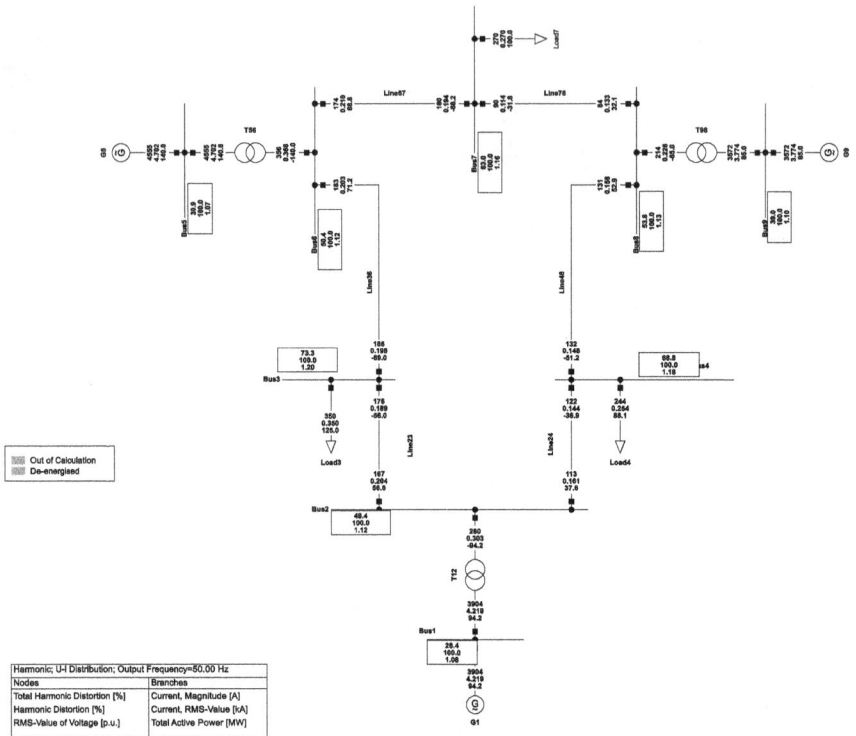

FIGURE 8.12
Harmonic load flow output of the 9-bus network.

8.6.2 Frequency Sweep

Step 1:
Right-click on Bus 4 and do as shown in Figure 8.14.

Step 2:
Choose the impedance variable to be the plottable variable, as shown in Figure 8.15. Most of the time, all the parameters are chosen; simply click "OK" and "Close."

Step 3:
Create a new plot page. See Figure 8.16.

Step 4:
Create a new curve plot. See Figure 8.17.

Step 5:
Run calculate impedance frequency characteristics. See Figure 8.18.

Step 6:
Double-click on the graphic white screen and act as shown in Figure 8.19.

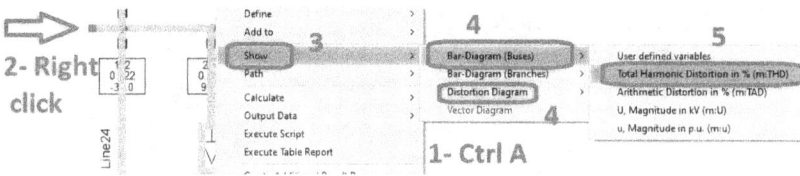

FIGURE 8.13
Draw the THD diagram.

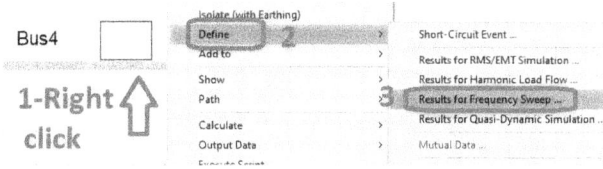

FIGURE 8.14
Results for frequency sweep.

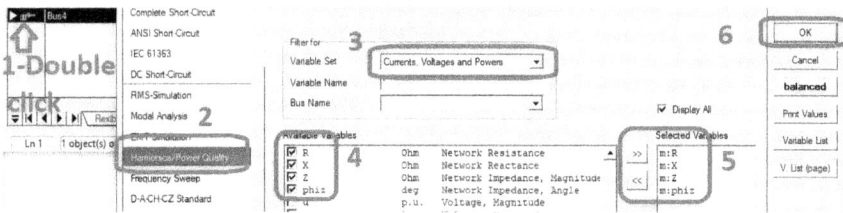

FIGURE 8.15
Select impedance as a graph variable.

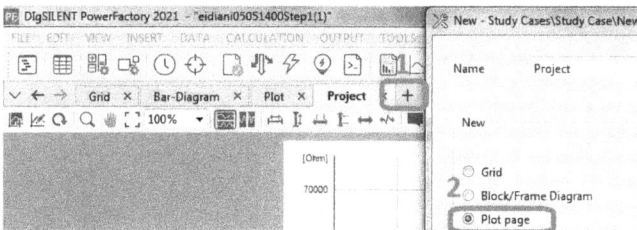

FIGURE 8.16
Create a new plot page.

Step 7:

Run calculate impedance frequency characteristics again. See Figure 8.20.

Step 8:

Determine the precise frequency and harmonic order value by zooming in on the peak point and choosing it, as shown in Figure 8.21.

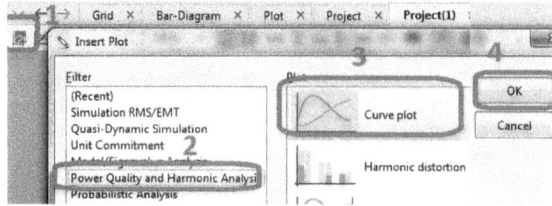

FIGURE 8.17
Create a new curve plot.

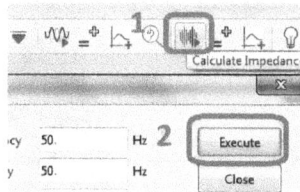

FIGURE 8.18
Calculate impedance frequency characteristics.

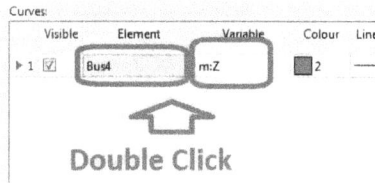

FIGURE 8.19
Determine variable and bus.

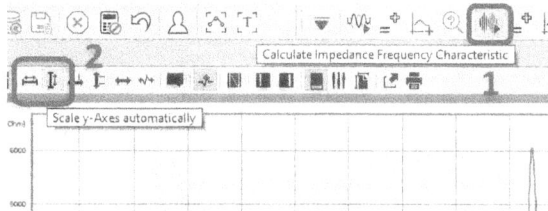

FIGURE 8.20
Calculate impedance frequency characteristics and scale axes automatically.

Step 9:

The impact of altering the harmonic model of loads in harmonic analysis is examined in this section. There are multiple harmonic models for loads. How to select the harmonic model for load 3 is illustrated in Figure 8.22. Compare the response to the case without the harmonic model after doing this for all loads.

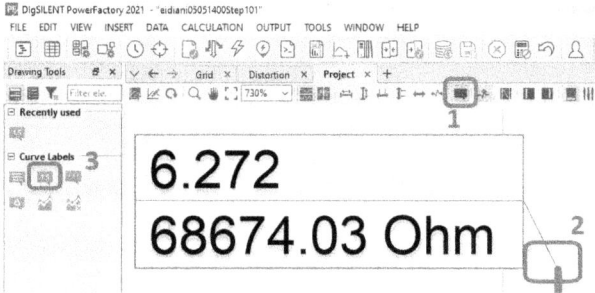

FIGURE 8.21
Determine the precise frequency and harmonic order value.

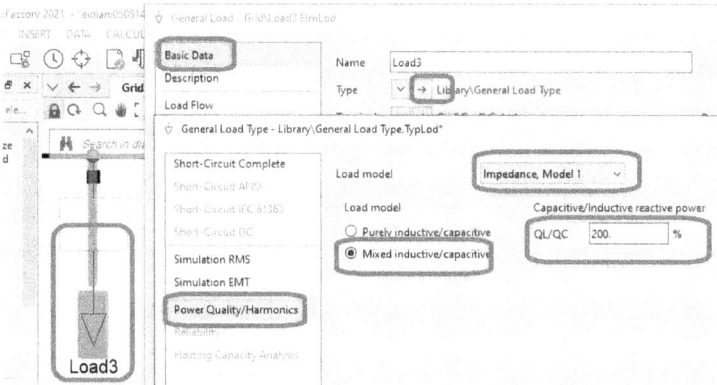

FIGURE 8.22
Different harmonic models for load.

8.7 Summary

Power systems often experience deviations from ideal sinusoidal voltage and current waveforms due to factors like transformer core saturation and power electronic converters, leading to the generation of harmonics—integer multiples of the fundamental frequency. While odd harmonics are common, even harmonics are rare in the symmetrical networks. This chapter discusses the characteristics of these harmonics, particularly in three-phase systems, and introduces DIgSILENT PowerFactory software for analyzing harmonic networks, with additional resources provided for further exploration.

References

1. https://mcelabs.com/powerqualityteachingtoy.
2. https://www.woodlandscoalition.com/PDF's/KEMAHarmonic AnalysisReport.pdf
3. Saenz, R.A., *Introduction to Harmonic Analysis*, Student Mathematical Library, American Mathematical Society, 2023.
4. Arrillaga, J., Watson, N.R., *Power System Harmonics*, 2nd edition, Wiley, 2003.
5. Eidiani, M. "Integration of renewable energy sources," In: Fathi, M., Zio, E., Pardalos, P.M. (eds) *Handbook of Smart Energy Systems*. Springer, Cham, 2022. https://doi.org/10.1007/978-3-030-72322-4_41-1
6. Eidiani, M., Zeynal, H., Ghavami, A., Zakaria, Z. "Comparative analysis of mono-facial and bifacial photovoltaic modules for practical grid-connected solar power plant using PVsyst," *2022 IEEE International Conference on Power and Energy (PECon)*, 2022, pp. 499–504, https://doi.org/10.1109/PECon54459.2022.9988872
7. Eidiani, M., Zeynal, H., Zakaria, Z. "A comprehensive study on the renewable energy integration using DIgSILENT," *2023 IEEE 3rd International Conference in Power Engineering Applications (ICPEA)*, Putrajaya, Malaysia, 2023, pp. 197–201, https://doi.org/10.1109/ICPEA56918.2023.10093153
8. Parhamfar, M., Eidiani, M., Abtahi, M. "Distributed energy storage system: Case study," In: Hossain E., Ghosh, A. (eds) *Distributed Energy Storage Systems for Digital Power Systems*. Springer, Cham, 2024, pp. 395–422.
9. Ahmadi, M., Mousavi, M. H., Moradi, H., Rouzbehi, K., "A new approach for harmonic detection based on eliminating oscillatory coupling effects in microgrids," *IET Renewable Power Generation*, vol. 17, no. 14, pp. 3536–3553, 2023.

9

Optimal Capacitor Placement Methods

9.1 Introduction

One of the key aspects of designing distribution networks is the optimal placement of shunt capacitors to compensate for the reactive current demand of loads and lines at various load levels. Distribution networks experience substantially higher power losses than transmission networks, primarily due to their radial topology and electrical characteristics. These networks have resistance-to-reactance (R/X) ratios often exceeding 0.5, compared to transmission networks' typical R/X ratios below 0.2. The radial structure necessitates power flow through sequential line segments, while the higher R/X ratio results in increased thermal losses and voltage regulation challenges. These inherent characteristics make distribution networks responsible for approximately 70% of total system losses in typical power systems. Installing capacitors in distribution networks can significantly reduce voltage drops, free up capacity on transmission lines, and minimize losses. However, economic factors, such as the cost of capacitors, installation, and the potential financial benefits, influence the decision to install capacitors.

Capacitors used in distribution networks are categorized into three types based on their connection to the network: permanent, non-permanent, and controlled capacitors.

Permanent capacitors are designed to handle the base load of the network and remain connected continuously. Non-permanent capacitors are connected during peak load periods to compensate for additional reactive power demand. Controlled capacitors are automatically switched on and off based on voltage, current, or power factor levels to optimize reactive power compensation.

The objective function for optimizing capacitor placement and sizing typically focuses on minimizing network losses. Constraints such as bus voltage limits, installation limitations, and cost considerations must also be taken into account.

DOI: 10.1201/9781003590514-9

This chapter provides a brief overview of optimal capacitor placement methods and demonstrates the application of a method to determine the optimal location for capacitors on a simple network using DIgSILENT PowerFactory software. How to use the software and the possibility for further work are explained in the appendix. For more detailed work and further work, you can refer to the references [1–7] at the end of this chapter.

9.2 Economic Benefits of Capacitor Installation

Power capacitors are a cost-effective way to provide reactive power to a power system. When generators supply reactive power, it puts a strain on power system equipment like generators, transformers, transmission and distribution lines, and protective devices. They need to be larger to handle the increased load. Capacitors help by reducing the amount of reactive power that needs to be transmitted through power lines. This reduces power losses and current flow in lines and transformers. The more power factor is corrected with capacitors, the more capacity can be added to generators and substations. This improves their ability to handle additional load and maintain voltage regulation.

Here are the economic benefits of installing power capacitors:

From System Capacity Benefits:
- Increasing power transfer capability across transmission lines, substations, and feeders
- Improving overall system capacity utilization

From Power Quality Benefits:
- Reducing voltage drop and improving voltage regulation.
- Reducing power losses (copper losses).
- Increasing power factor at installation points and source generators

From Financial Benefits:
- Reducing power demand and associated costs.
- Postponing or eliminating the need for system upgrades and new equipment.
- Reducing overall investment costs per kilowatt of delivered power.

9.3 Basics and Criteria for Selecting the Capacitor Installation Location

The installation of capacitor banks requires careful consideration through both technical and economic lenses. The decision-making process involves a comprehensive cost-benefit analysis that weighs potential advantages against the total investment required. The total cost encompasses not just the initial procurement of equipment, but also installation, commissioning, ongoing maintenance, and potential repairs of both the capacitors and their associated feeder systems.

Location plays a crucial role in maximizing benefits—the closer capacitors are installed to the point of consumption, the greater their positive impact on the overall system. This proximity principle helps optimize power factor correction and minimize transmission losses.

Voltage level selection represents another critical factor, particularly from a financial perspective. While capacitors are available across a wide range of voltages, the optimal cost-effectiveness typically occurs in the 6–20 kV range. Although higher voltage capacitors exist, their substantially higher costs often make them less economically viable unless specifically required by system conditions.

The investment decision ultimately hinges on whether the cumulative benefits—including improved system capacity, reduced losses, and enhanced voltage regulation—exceed the total lifecycle costs of the installation. This evaluation should consider both immediate advantages and long-term operational benefits to ensure a sound investment decision.

9.4 Optimal Capacitor Placement Methods

Here are some popular and novel approaches, along with their advantages and disadvantages (see Table 9.1). The best location for the capacitor should take into account factors like the capacitor's size, the financial advantages of doing so, system constraints, and environmental impacts like electromagnetic interference and noise pollution.

9.5 Supercapacitors

Supercapacitors represent a transformative technology in power system applications, particularly for reactive power compensation and

TABLE 9.1

Optimal Capacitor Placement Methods

Method		
Advantages	**Disadvantages**	**Examples**
Metaheuristic Optimization Techniques		
Near-optimal solutions efficiently, Handle complex, non-linear, and non-convex optimization problems. Flexible and adaptable to various system constraints.	Not guarantee global optimality. Computational complexity can increase with larger systems.	• Genetic Algorithms (GA) • Particle Swarm Optimization (PSO) • Differential Evolution (DE) • Simulated Annealing (SA) • Grey Wolf Optimizer (GWO)
Machine Learning and Artificial Intelligence		
Learn from historical data and adapt to changing system conditions. Handle large-scale and complex systems. Identify patterns and trends that may not be obvious to human experts.	Requires large amounts of high-quality data. Can be computationally expensive. May be difficult to interpret and explain.	• Neural Networks • Support Vector Machines (SVM) • Reinforcement Learning • Deep Learning
Hybrid Methods		
Combine the strengths of multiple methods. Can improve accuracy and efficiency. Can handle complex and dynamic systems.	Can be more complex to implement. May require significant computational resources.	• Combining metaheuristic algorithms with machine learning techniques. • Combining power flow analysis with sensitivity analysis.
Multi-Objective Optimization		
Can consider multiple objectives simultaneously, such as minimizing power losses, improving voltage profile, and reducing investment costs. Can provide a more comprehensive solution.	Can be more complex to solve. May require more computational resources.	• Weighted Sum Method • ε-Constraint Method • Goal Programming • Evolutionary Algorithms • Multi-Objective Particle Swarm Optimization (MOPSO)
Real-Time Optimization		
Can adapt to changes in system conditions in real time. Can improve system performance and reliability.	Requires fast and accurate data acquisition and processing. Can be computationally demanding.	• Model Predictive Control (MPC) • Reinforcement Learning (RL) • Online Optimization Techniques

(Continued)

TABLE 9.1 (*Continued*)

Optimal Capacitor Placement Methods

Method		
Advantages	**Disadvantages**	**Examples**
Sensitivity Analysis		
Identifies buses that are most sensitive to changes in reactive power. Provides a quick and efficient way to identify potential capacitor placement locations.	May not consider the overall system impact of capacitor placement. Requires accurate system data.	• Power Flow Analysis • Jacobian Matrix Analysis • Admittance Matrix Analysis • Contingency Analysis

voltage stability. These devices, also known as ultracapacitors, occupy a unique position between conventional capacitors and batteries, combining rapid response capabilities with significant power density advantages.

The fundamental operating principle of supercapacitors relies on electrostatic charge storage through an electric double-layer mechanism. When integrated into power systems, this characteristic enables them to respond to voltage fluctuations and power quality issues within milliseconds—far faster than traditional reactive power compensation methods. Their electrode-electrolyte interface creates a massive effective surface area, allowing for substantial charge storage capacity while maintaining rapid charge-discharge capabilities.

In power system applications, supercapacitors excel in several critical areas:

- **Dynamic Voltage Support**: Their rapid response time makes them ideal for managing voltage sags and swells in distribution networks.

- **Reactive Power Compensation**: They can provide instantaneous reactive power support, helping maintain power factor and system stability.

- **Transient Stability Enhancement**: Their ability to absorb or inject power quickly helps dampen power system oscillations.

- **Ride-Through Support**: During brief power interruptions, they can maintain critical loads while primary power sources recover.

However, implementation considerations must account for certain technical constraints. The relatively lower energy density compared to batteries means they are better suited for short-term power quality improvements rather than long-term energy storage. Additionally, their self-discharge characteristics and voltage-dependent capacity require sophisticated control systems for optimal integration into power networks.

The most effective applications of supercapacitors in modern power systems typically involve hybrid configurations, where they complement other power quality improvement devices. For instance, when paired with static VAR compensators or STATCOM systems, supercapacitors can enhance the overall system response to power quality events while providing crucial backup during switching operations.

As power systems continue to evolve with increasing renewable energy integration and growing power quality demands, supercapacitors' role in reactive power compensation and system stability is becoming increasingly vital. Their continued development, particularly in terms of energy density improvements and cost reduction, promises to further expand their utility in maintaining reliable and efficient power system operation.

9.6 Optimum Capacitor Placement with DIgSILENT

This section demonstrates how to use DIgSILENT PowerFactory software to simulate a basic harmonic network without any prior knowledge of the program. You can read and follow the appendix section of this chapter to repeat these simulations.

First, a network without a capacitor and with overload is taken into consideration for the analysis. A 9-bus network with overload and a 6.86 MW loss is shown in Figure 9.1. (Download the file (Chapter9-1.pfd))

Please be aware that the value, cost, and quantity of capacitors in the warehouse all affect the optimum capacitor placement solution. Figure 9.2 provides an example of the data that is available for analysis. Whether there is a penalty factor for voltage deviation from 1 pu is the next crucial question. Furthermore, does exceeding the acceptable minimum and maximum voltage range (between 0.95 and 1.05) result in an extra penalty factor? See Figure 9.3. Ultimately, you should be aware of the best way the optimal capacitor placement, which involves using DIgSILENT software's Optimization and Sensitivity Analysis features. In addition, what is the energy cost $/kWh? The DIgSILENT software is used to implement the optimal capacitor placement, which results in the placement of the capacitors in the network and their display in Figures 9.4 and 9.5 graphic outputs.

The two versions, v15.1 and v2021, have different network outputs. The text output of these two editions in Figures 9.6 and 9.7 makes this difference evident for easier comprehension. Although the network losses have decreased and more capacitors have been added in the v2021, the bus voltages in the v2021 have surpassed the acceptable voltage range. In fact,

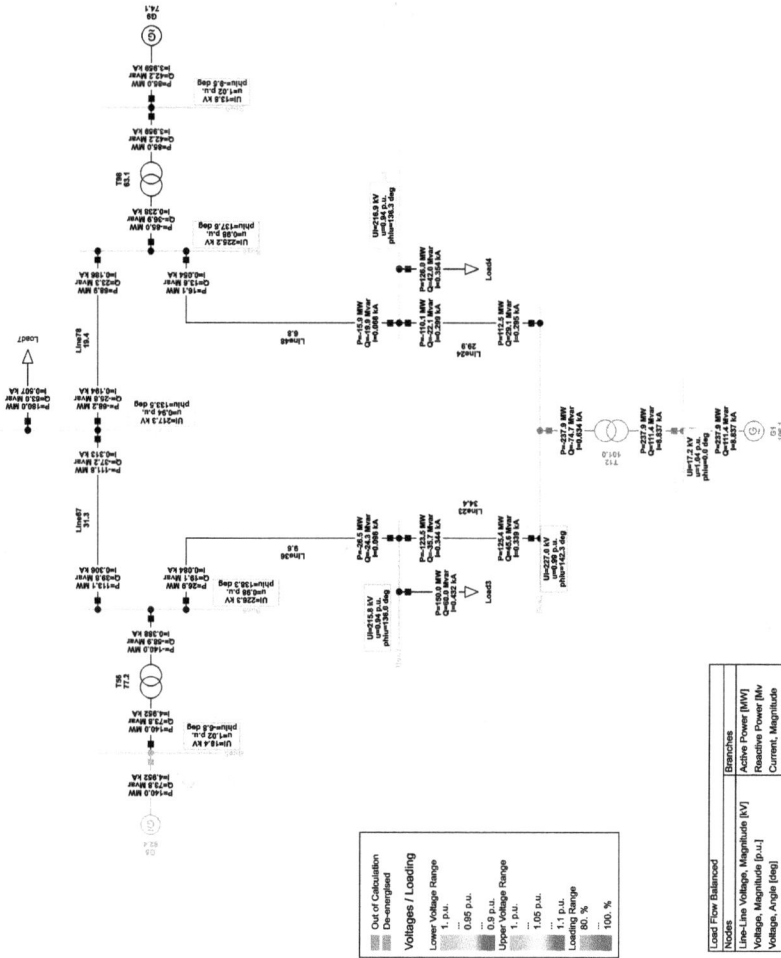

FIGURE 9.1
The output of the 9-bus system with overload (v2021).

Load Flow Balanced	
Nodes	Branches
Line-Line Voltage, Magnitude [kV]	Active Power [MW]
Voltage, Magnitude [p.u.]	Reactive Power [Mv
Voltage, Angle [deg]	Current, Magnitude

Out of Calculation
De-energised

Voltages / Loading

Lower Voltage Range
1. p.u.
0.95 p.u.
0.9 p.u.

Upper Voltage Range
1. p.u.
1.05 p.u.
1.1 p.u.

Loading Range
80. %
100. %

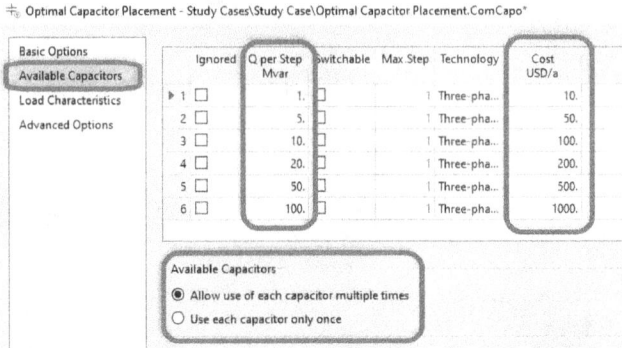

FIGURE 9.2
Capacitor bank information available in the warehouse.

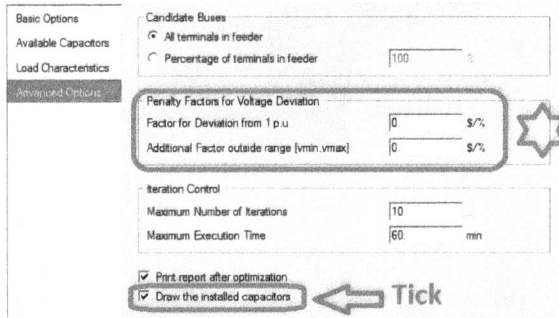

FIGURE 9.3
Penalty factors for voltage deviation.

the v2021's response can be made identical to the v15.1 by including the penalty factor for deviation from the acceptable voltage range.

The purpose of this comparison is to demonstrate that the results differ for the same network, similar conditions, identical software, and two distinct versions. Note that both of the responses are accurate. Analysis should only be done on outputs.

DIgSILENT applies the following steps to find the optimal configuration of capacitors:

a. **Optimisation method:**
- The "best" candidate terminal is first identified through a sensitivity analysis, which connects the largest capacitor from the user-defined list of capacitors to each target feeder terminal in order to assess the effect on the overall cost (Losses + Voltage Violations). The price of the biggest capacitor is not included at this time.

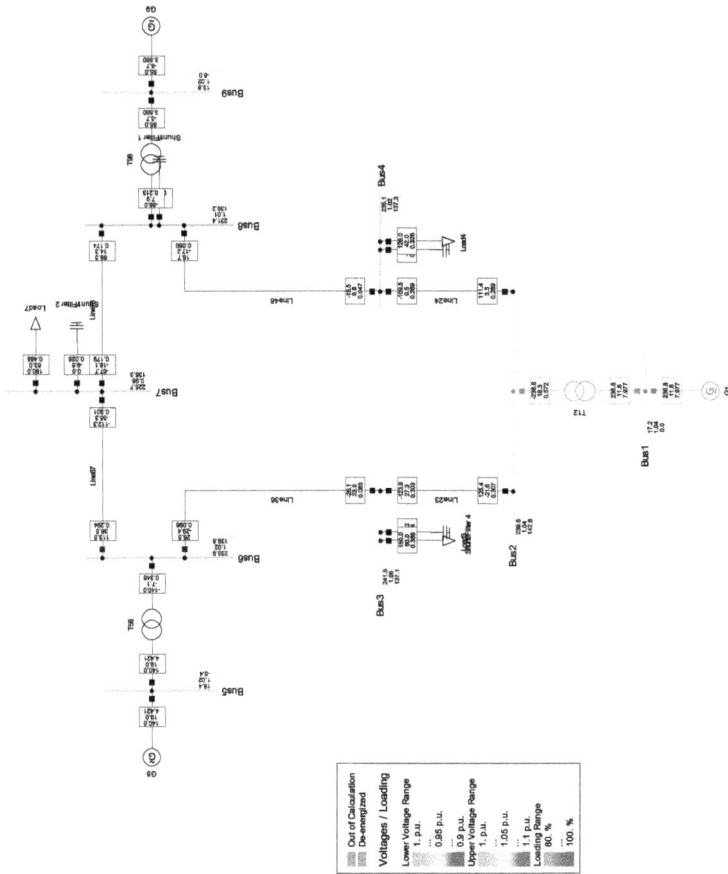

FIGURE 9.4
Graphical network output after adding capacitors to improve the network (loss reduction and voltage improvement) v15.1.

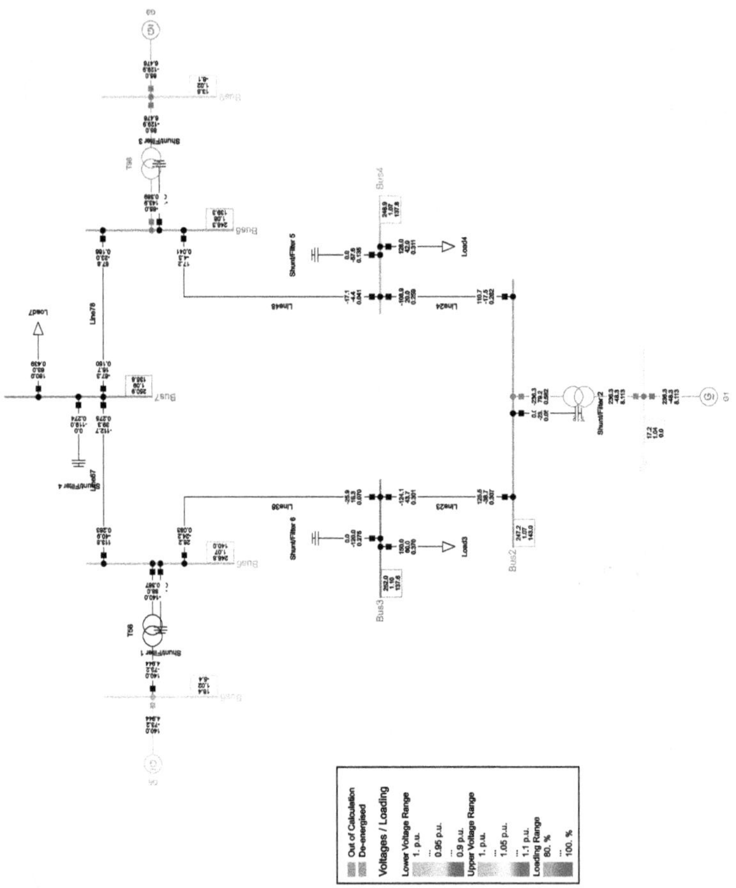

FIGURE 9.5
Graphical network output after adding capacitors to improve the network (loss reduction and voltage improvement) v2021.

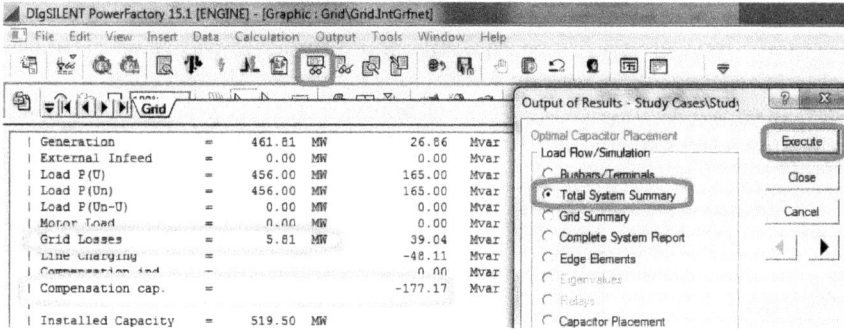

FIGURE 9.6
Text network output after adding capacitors to improve the network (loss reduction and voltage improvement) v15.1.

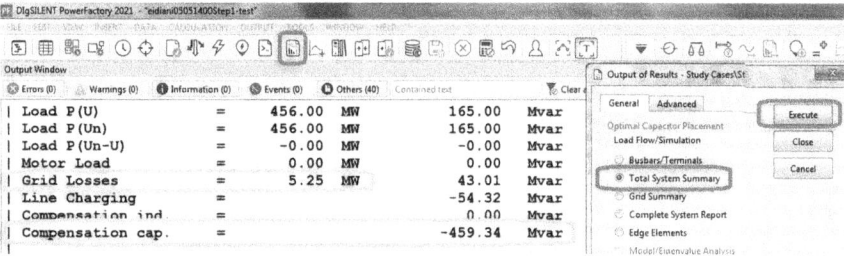

FIGURE 9.7
Text network output after adding capacitors to improve the network (loss reduction and voltage improvement) v2021.

- In terms of overall cost savings, terminals are arranged in descending order. The "best" candidate terminal for a "new" capacitor is the one that offers the biggest cost savings.
- The optimization process then uses each available capacitor from the user-defined list, including the cost of each capacitor, to assess the cost reduction at the candidate terminal. When taking into account the capacitor's yearly cost, the "best" capacitor is the one that offers the biggest cost savings.
- Step 1 should be repeated, but this time the ranking of candidate terminals excludes any terminals that have already been chosen as potential candidates for capacitor installation. When all terminals have capacitors installed or when installing capacitors is no longer able to lower costs, the algorithm comes to a halt.

b. **Sensitivity Analysis Method:**
Using the sensitivity analysis method, the candidate terminals are ranked based on how they affect the overall loss cost, which does not include the cost of the capacitor. The output window displays the output. With this option, one can quickly see where a single capacitor would work best. If this option is chosen, there are no capacitors installed.

How to use the software and the possibility for further work are explained in the appendix. For more detailed work and further work, you can refer to the references [1–7] at the end of this chapter.

9.7 Two-Choice Questions (Yes/No)

1. Is optimal capacitor placement used to improve power system efficiency and reliability?
2. Can capacitors reduce power losses in transmission and distribution systems?
3. Does capacitor placement contribute to power factor correction?
4. Can capacitor placement improve the voltage profile in power systems?
5. Does over-capacitation lead to voltage rise and instability in power systems?
6. Is sensitivity analysis used to identify the most suitable buses for capacitor placement?
7. Is power flow analysis a tool for evaluating the impact of capacitor placement?
8. Can metaheuristic algorithms, like genetic algorithms and particle swarm optimization, be used for optimal capacitor placement?
9. Are artificial intelligence techniques used to optimize capacitor placement in complex power systems?
10. Do the economic benefits of capacitor placement always outweigh the initial investment costs?
11. Can capacitor placement help defer or avoid investments in new power system equipment?
12. Is the optimal capacitor placement solution always unique?

13. Can real-time optimization techniques adjust capacitor settings based on system conditions?

14. Can capacitor placement mitigate the impact of renewable energy integration on power system stability?

15. Is the size of capacitors determined based on load characteristics and system requirements?

16. Are capacitor banks typically switched on and off manually?

17. Are power factor correction capacitors usually connected in series with the load?

18. Can capacitor banks improve the damping of power system oscillations?

19. Does the installation of capacitors sometimes increase power system losses?

20. Can proper capacitor placement mitigate power quality issues like voltage fluctuations and harmonics?

21. Is the cost of capacitors and their installation a significant factor in decision-making?

22. Should environmental factors like noise pollution and electromagnetic interference be considered in capacitor placement?

23. Can capacitor banks improve the efficiency of electric motors?

24. Can the optimal capacitor placement solution change over time due to load or system changes?

25. Are power system simulations essential for evaluating the impact of capacitor placement?

26. Does a higher power factor reduce the apparent power required for delivering real power?

27. Can capacitor banks reduce voltage drops in distribution systems?

28. Is the reactive power demand of a power system constant throughout the day?

29. Can capacitor banks improve power system stability during disturbances?

30. Is the optimal location for capacitor placement always at the load end of a feeder?

9.7.1 Key Answers to Two-Choice Questions

Yes	Other
No	10, 12, 16, 17, 28, 30

9.8 Appendix, Optimum Capacitor Placement in DIgSILENT

This section requires a basic understanding of DIgSILENT PowerFactory software. To proceed with this section, you need to download the file (Chapter9.pfd) from the book's end attachments. Proceed with the remainder of this step after importing the network. It is recommended to do the following steps.

Step 1:
First, run the software and by running **Load flow,** you will get to Figure 9.8.

Step 2:
On the text output screen (Figure 9.9), note the losses (No. 5) and the installed capacitor (No. 6).

Step 3:
In this step, a **Feeder** must be defined. Right-click on line23 key near bus 2 (as shown in Figure 9.10), define a Feeder, and click Ok.

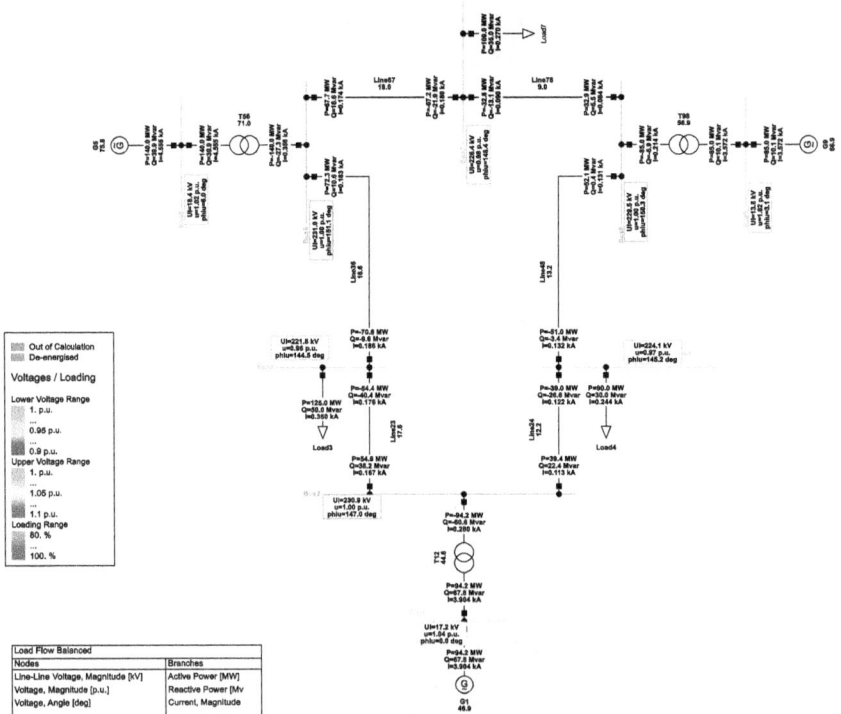

FIGURE 9.8
The graphical output of the 9-bus system under normal conditions (v2021).

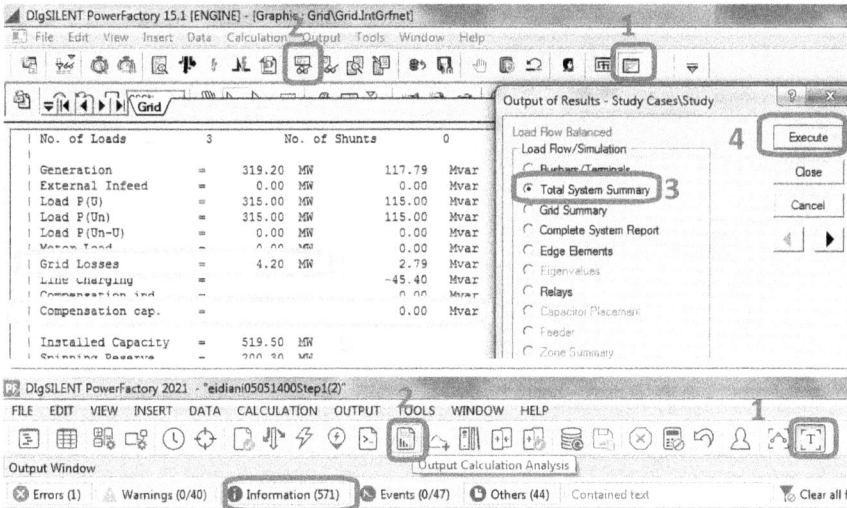

FIGURE 9.9
The text output of the 9-bus system under normal conditions (v2021 and v15.1).

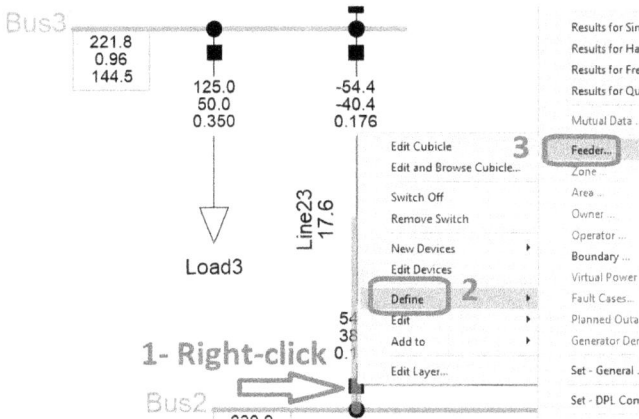

FIGURE 9.10
Define a Feeder.

Step 4:

Run the load flow. Go to the software's colouring section and switch the colouring from Low and High Voltage to Feeders to guarantee the feeder's definition, as seen in Figure 9.11. The figure needs to be painted all the way through. As a result, the feeder is specified for the whole network. The feeder was not properly chosen if it was not in full color. You must remove it and repeat the next step. Now switch back to the Low and High Voltage coloring mode.

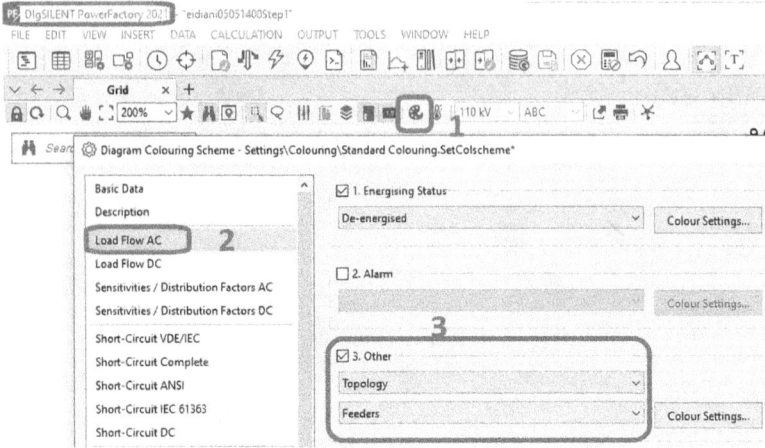

FIGURE 9.11
Topology colouring based on Feeder.

Step 5:

In order to achieve overload and voltage reduction in the network and to better observe the capacitor effect, the loads have been increased in this step. As shown in Figure 9.12, you should raise the loads' scale factor to ensure that their active and reactive power increases proportionately.

If you run load flow, the network will be overloaded and under-voltage, and a 6.86 MW loss as shown in Figure 9.1.

Step 6:

As in Figure 9.13, from the ToolBox icon (No. 1), select **Optimal Capacitor Placement** or **Distribution Network Optimisation** (No. 2) and click **Capacitor Calculations** (No. 3).

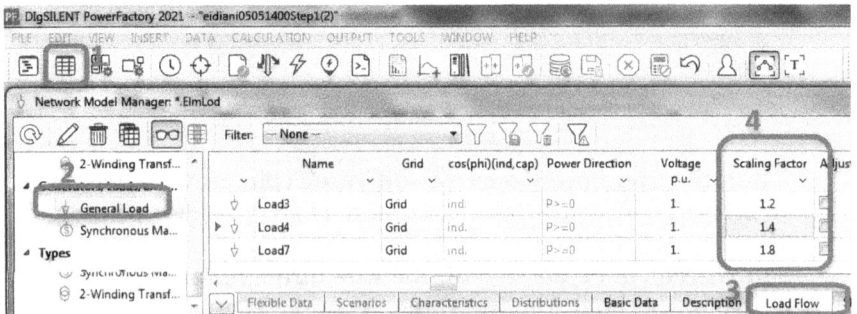

FIGURE 9.12
Load increase with scale factor.

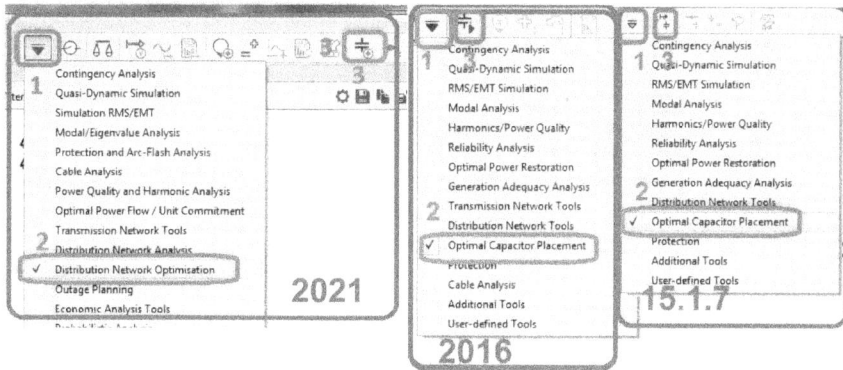

FIGURE 9.13
Changing the location and name of the capacitor optimal placement toolbox in different versions of DIgSILENT.

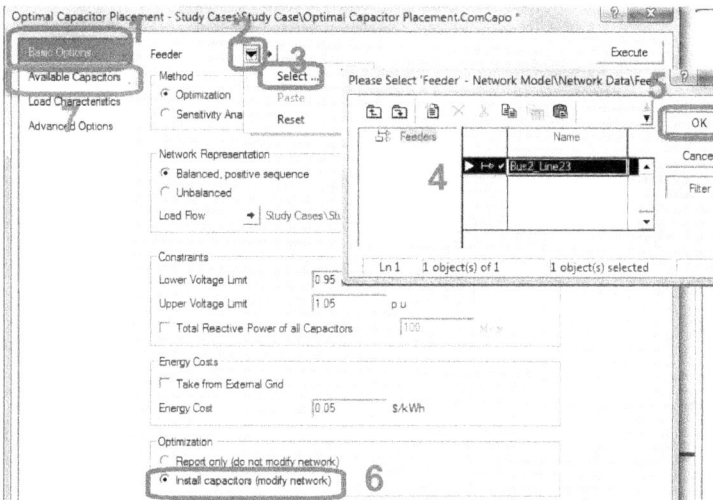

FIGURE 9.14
Optimal capacitor placement window.

Step 7:

In the window shown in Figure 9.14, first select the feeder according to numbers 2–5. Then check (No. 6) to add the capacitors to the network. At the end, tick (No. 7) to go to the section (**Available Capacitor**) (next step) and define the capacitor bank.

Step 8:

Any number of capacitors can be added to the bank in the **Available Capacitors** bank section. Five capacitors of varying values are added in this step. Capacitors are added in Figure 9.2. Press Advanced Options in the figure at the end to move on to the next step.

Step 9:

To ensure that the best capacitors are connected to the network, activate the options in Figure 9.3 by going to the Advanced Options after the previous step. Finally, press the Execute key and then the OK key multiple times.

Step 10:

Figures 9.4 and 9.5 display the network's output following the addition of ideal capacitors under identical conditions for a network in v15.1 and v2021, but the results differ! Why?

Step 11:

Figures 9.6 and 9.7 show the text output of the installed capacitors in two versions that are different.

Step 12:

To view the installed capacitor's details, follow the steps shown in Figure 9.15. Which buses are added, and how many capacitors are there?

Step 13:

You can delete all of the added capacitors and get back to the main network by pressing the (💬) or (🗨) key (Remove Previous Solution). You can also do the capacitor again by pressing the (▶╬) or (╬) key (Calculate Optimal Capacitor Placement).

Step 14:

Right-click on a line and draw the voltage profile. Compare these figures for two cases with capacitors (Figure 9.16) and without capacitors (Figure 9.17).

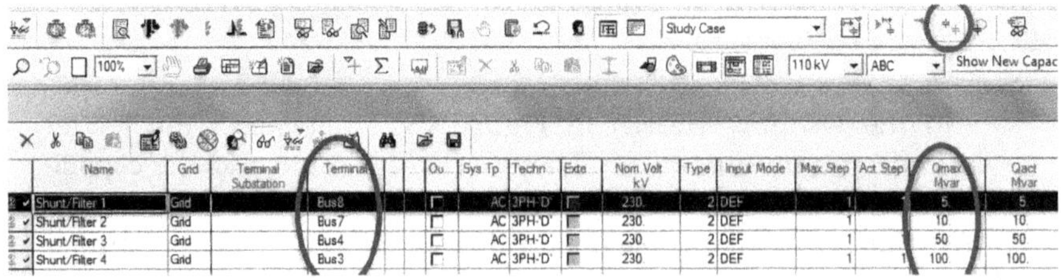

FIGURE 9.15
Optimal capacitor final placement window.

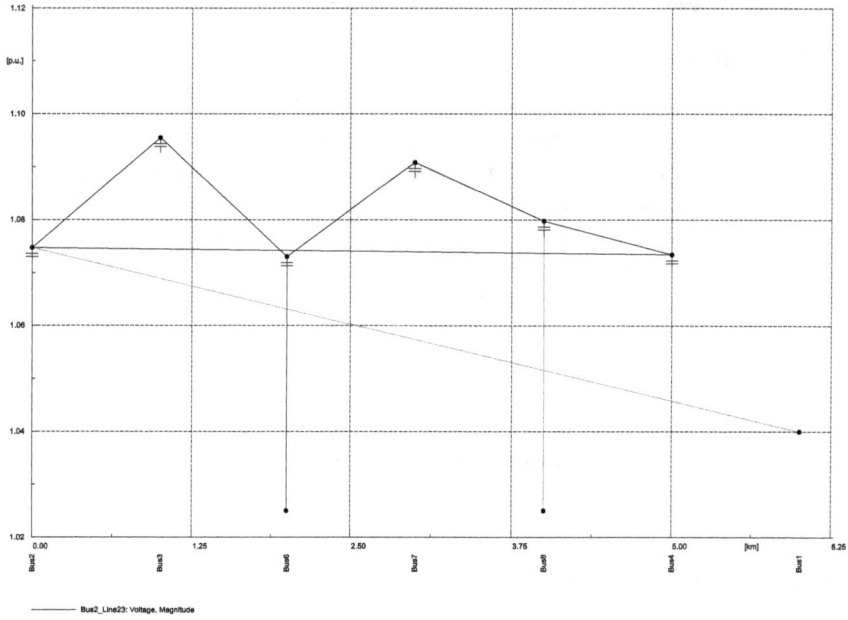

FIGURE 9.16
Voltage profile with optimal capacitor.

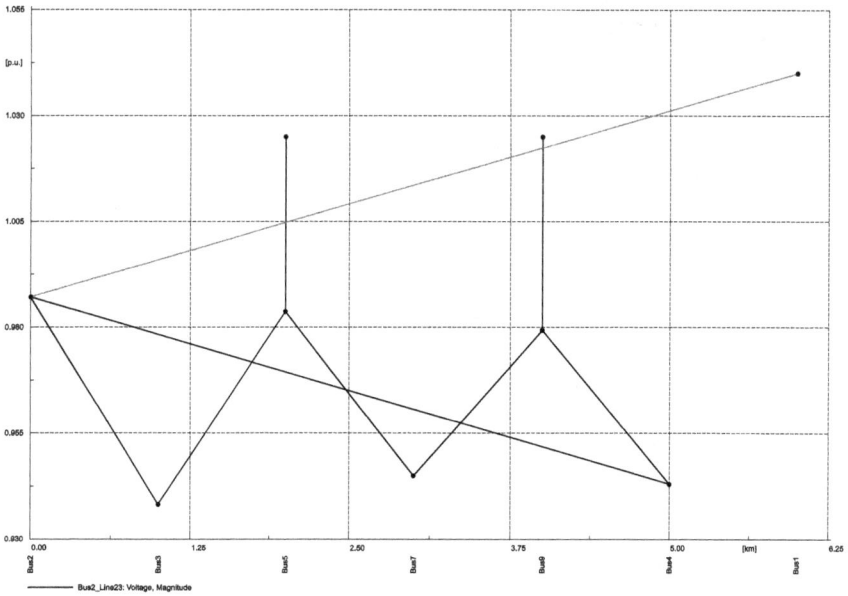

FIGURE 9.17
Voltage profile without capacitor.

9.9 Summary

The design of distribution networks involves strategically placing shunt capacitors to manage reactive current demands, which helps reduce losses and voltage drops in these networks. Distribution networks, characterized by their radial structure and high resistance-to-reactance ratio, are particularly prone to losses, making capacitor installation a valuable solution. Capacitors are classified into permanent, non-permanent, and controlled types, each serving different roles in reactive power compensation. The optimization of capacitor placement focuses on minimizing network losses while considering economic factors and operational constraints. This chapter outlined methods for optimal capacitor placement and demonstrated their application using DIgSILENT PowerFactory software, with additional resources provided for further exploration.

References

1. Zeynal, H., Jiazhen, Y., Azzopardi, B., Eidiani, M. "Impact of Electric Vehicle's integration into the economic VAr dispatch algorithm," *2014 IEEE Innovative Smart Grid Technologies—Asia (ISGT ASIA)*, 2014, pp. 780–785, https://doi.org/10.1109/ISGT-Asia.2014.6873892
2. Eidiani, M., Ashkhane, Y., Khederzadeh, M. "Reactive power compensation in order to improve static voltage stability in a network with wind generation," *2009 International Conference on Computer and Electrical Engineering, ICCEE 2009*, 2009, 1, pp. 47–50, 5380672, https://doi.org/10.1109/ICCEE.2009.239
3. Eidiani, M., Heidari, V., *Fundamentals of Power Systems Analysis 1: Problems and Solutions*, Taylor & Francis Group, CRC Press, 2023, pp. 1–215, https://doi.org/10.1201/9781003394433
4. Eidiani, M., Rouzbehi, K., *Fundamentals of Power System Transformers Modeling, Analysis, and Operation*, Taylor & Francis Group, CRC Press, 2025, pp. 1–127. http://www.routledge.com/9781032881751
5. Momen, S., Hekmati, A., Majidi, S., Zand, Z., Zand, M., Nikoukar, J., Eidiani, M. "Energy harvesting for smart energy systems," In: Javadi, M.S., Godina, R., Rodrigues, E. (eds) *Handbook of Smart Energy Systems*. Springer, Cham, 2022. https://doi.org/10.1007/978-3-030-72322-4_12-1
6. Eidiani, M. "Modeling renewable energy resources using DIgSILENT PowerFactory software," In: Vasant, P., Weber, G.W. (eds) *Power Systems Operation with 100% Renewable Energy Sources*. Springer, Cham, 2023, pp. 165–202.
7. Eidiani, M., Zeynal, H., Zakaria, Z. "A comprehensive study on the renewable energy integration using DIgSILENT," *2023 IEEE 3rd International Conference in Power Engineering Applications (ICPEA)*, Putrajaya, Malaysia, 2023, pp. 197–201, https://doi.org/10.1109/ICPEA56918.2023.10093153

10

Asymmetrical Distribution Networks

10.1 Introduction

Asymmetrical conditions in distribution networks, caused by unbalanced loads and line configurations, present significant challenges to power system engineers. This imbalance can lead to increased power losses, voltage imbalances, and potential equipment damage. Unbalanced loads occur when the current flowing through each phase of a three-phase system is unequal in magnitude or phase angle. This can be caused by factors such as unequal phase loading, single-phase loads, and faulty connections.

Unbalanced lines occur when the impedance of each phase in a three-phase system is unequal. This can be caused by unequal phase lengths, transposition errors, and unbalanced loads. The consequences of unbalanced loads and lines include increased power losses, voltage unbalance, equipment damage, reduced power system efficiency, and increased harmonic distortion.

To mitigate these issues, various techniques can be employed, including load balancing, phase shifting transformers, capacitor banks, and phase balancing transformers. By understanding the causes and consequences of unbalanced loads and lines, power system engineers can implement effective strategies to ensure the reliable and efficient operation of distribution networks.

This chapter begins with a review of the relationships needed to analyze asymmetrical and unbalanced networks. Next, the impact of unbalanced and asymmetrical loads and lines, as well as their combination, has been examined in a basic network. The final section explains how to use DIgSILENT software to simulate an asymmetrical and unbalanced network. For more detailed work and further work, you can refer to the references [1–7] at the end of this chapter.

10.2 Symmetrical Components[1]

Symmetrical components are a mathematical tool to simplify the analysis of unbalanced three-phase power systems. They break down unbalanced voltages and currents into three balanced sets of components: positive

DOI: 10.1201/9781003590514-10

sequence, negative sequence, and zero sequence. Positive sequence components represent a balanced three-phase system with a phase sequence of ABC. They are used to analyze normal, balanced operating conditions. Negative sequence components represent a balanced three-phase system with a phase sequence of ACB. They are used to analyze unbalanced conditions, such as single-phase faults and unbalanced loads.

Zero sequence components represent a balanced three-phase system with all three phases in phase. They are used to analyze ground faults and other conditions involving ground impedance. By breaking down unbalanced systems into these symmetrical components, engineers can analyze complex fault conditions and system disturbances more easily. This technique is widely used in power system analysis, fault analysis, and protection system design.

Consider the following definition:

$$\alpha \triangleq 1\angle 120° = -\frac{1}{2} + j\frac{\sqrt{3}}{2} \Rightarrow \alpha^2 = 1\angle 240° = \alpha* = -\frac{1}{2} - j\frac{\sqrt{3}}{2} \quad (10.1)$$

The relationship between voltage components can be summarized as the following equation.

$$\hat{V}_a^0 = \hat{V}_b^0 = \hat{V}_c^0, \quad \hat{V}_b^+ = \alpha^2 \hat{V}_a^+, \quad \hat{V}_b^- = \alpha \hat{V}_a^-, \quad \hat{V}_c^+ = \alpha \hat{V}_a^+, \quad \hat{V}_c^- = \alpha^2 \hat{V}_a^- \quad (10.2)$$

Equation (10.2) can also be shown in a matrix format as (10.3).

$$\begin{bmatrix} \hat{V}_a^0 \\ \hat{V}_a^+ \\ \hat{V}_a^- \end{bmatrix} = \frac{1}{3} \begin{bmatrix} 1 & 1 & 1 \\ 1 & \alpha & \alpha^2 \\ 1 & \alpha^2 & \alpha \end{bmatrix} \begin{bmatrix} \hat{V}_a \\ \hat{V}_b \\ \hat{V}_c \end{bmatrix},$$

$$\begin{bmatrix} \hat{V}_a \\ \hat{V}_b \\ \hat{V}_c \end{bmatrix} = \begin{bmatrix} 1 & 1 & 1 \\ 1 & \alpha^2 & \alpha \\ 1 & \alpha & \alpha^2 \end{bmatrix} \begin{bmatrix} \hat{V}_a^0 \\ \hat{V}_a^+ \\ \hat{V}_a^- \end{bmatrix} \quad (10.3)$$

Example 10.1

Find the symmetrical components of the following three-phase current [5].

$$\begin{cases} \hat{I}_a = 1.6\angle 25° \text{ pu} \\ \hat{I}_b = 1\angle 180° \text{ pu} \\ \hat{I}_c = 0.9\angle 132° \text{ pu} \end{cases}$$

Answer:

From equation (10.3):

$$
\begin{bmatrix} \hat{I}_a^0 \\ \hat{I}_a^+ \\ \hat{I}_a^- \end{bmatrix} = \frac{1}{3} \begin{bmatrix} 1 & 1 & 1 \\ 1 & \alpha & \alpha^2 \\ 1 & \alpha^2 & \alpha \end{bmatrix} \begin{bmatrix} 1.6\angle 25 \\ 1.0\angle 180 \\ 0.9\angle 132 \end{bmatrix}
$$

$$
= \frac{1}{3} \begin{bmatrix} 1.6\angle 25 - 1 + 0.9\angle 132 \\ 1.6\angle 25 - 1\angle 120 + 0.9\angle 12 \\ 1.6\angle 25 - 1\angle -120 + 0.9\angle 252 \end{bmatrix} = \begin{bmatrix} 0.4512\angle 96.4529° \\ 0.9435\angle -0.0550° \\ 0.6024\angle 22.3157° \end{bmatrix}
$$

$$
\begin{cases} \hat{I}_a^0 = \hat{I}_b^0 = \hat{I}_c^0 = 0.4512\angle 96.4529° \\ \hat{I}_b^+ = \alpha^2 \hat{I}_a^+ = 0.9435\angle -120.055° \quad ; \quad \hat{I}_c^+ = \alpha \hat{I}_a^+ = 0.9435\angle 119.945° \\ \hat{I}_b^- = \alpha \hat{I}_a^- = 0.6024\angle 142.3157° \quad ; \quad \hat{I}_c^- = \alpha^2 \hat{I}_a^- = 0.6024\angle -97.6843° \end{cases}
$$

Now the relationship between the components can be checked numerically.

$$
\begin{cases} \hat{I}_a = 1.6\angle 25 = \hat{I}_a^+ + \hat{I}_a^- + \hat{I}_a^0 \\ \quad = 0.9435\angle -0.0550 + 0.6024\angle 22.3157 + 0.4512\angle 96.4529 \\ \hat{I}_b = 1\angle 180 = \hat{I}_b^+ + \hat{I}_b^- + \hat{I}_b^0 \\ \quad = 0.9435\angle -120.055 + 0.6024\angle 142.3157 + 0.4512\angle 96.4529 \\ \hat{I}_c = 0.9\angle 132 = \hat{I}_c^+ + \hat{I}_c^- + \hat{I}_c^0 \\ \quad = 0.9435\angle 119.945 + 0.6024\angle -97.6843 + 0.4512\angle 96.4529 \end{cases}
$$

The analysis of an unbalanced network requires decomposing it into its positive, negative, and zero sequence components. Then, the superposition of these individual analyses yields the complete system behavior.

10.3 Asymmetrical Network Analysis

This section examines an asymmetrical network, as depicted in Figure 10.1. The rest of this section shows more detailed information by zooming in on each component of the network. Two subsystems, System1 and System2,

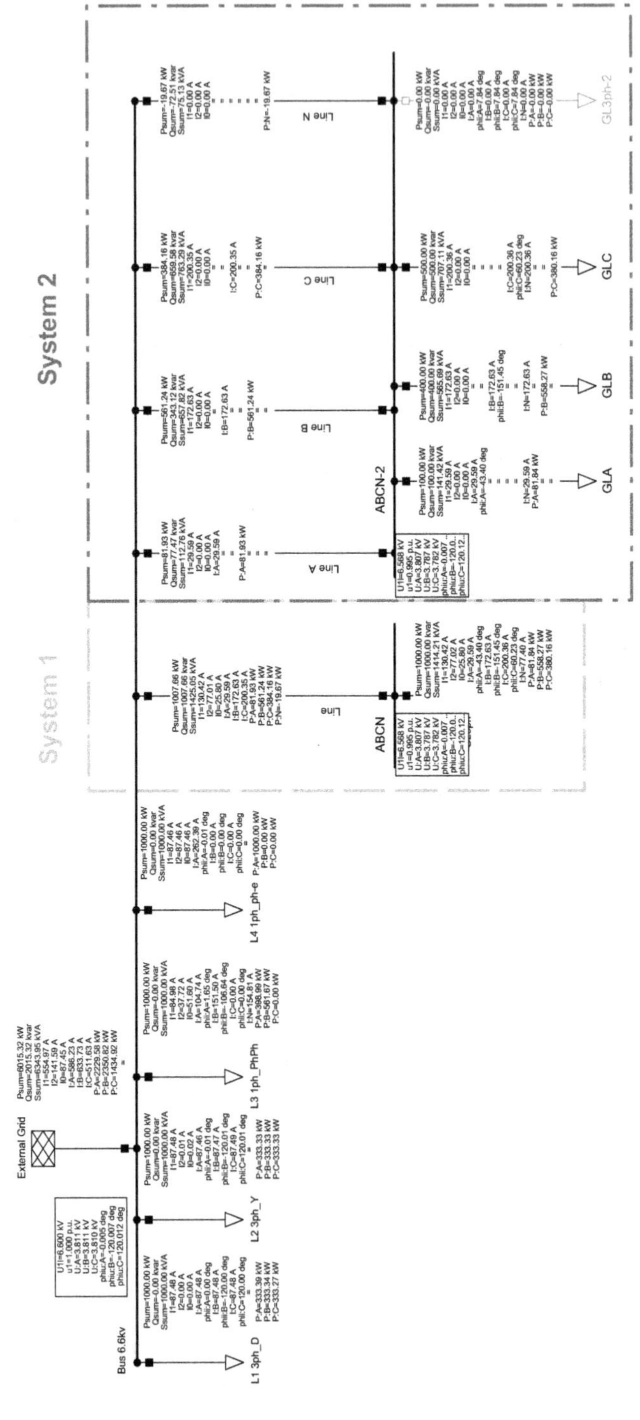

FIGURE 10.1

Analysis of a simple asymmetrical network.

and the main section of Figure 10.1, which has four loads, are examined and contrasted in the sections that follow. You can use this chapter's appendix to simulate this network.

Figure 10.2 shows the main part of Figure 10.1. In this system, 4 loads of 1 MW and 0 MVAr are used, and two systems 1 and 2 are disconnected. In Table 10.1, the characteristics of the 4 types of loads are specified.

The following points are evident when we examine the data in Figure 10.2.

1. Star and delta loads are similar in constant power. By comparing loads 1 (L1 3ph_D) and 2 (L2 3ph_Y), it can be seen that while converting a star load to a delta and vice versa alters the load's power consumption, the constant power of the two loads is the same. The source provides a steady, balanced current to these two symmetrical 1 MW loads, whether they are in star or delta configuration. A calculation error is the cause of the slight discrepancy.

$$P_{3ph} = |S_{3ph}| = \sqrt{3}(V_L)(I) \Rightarrow 1\,\text{MW} = \sqrt{3}(6.6\,\text{kV})(I_A) \Rightarrow I_A = 87.48\,\text{A}$$

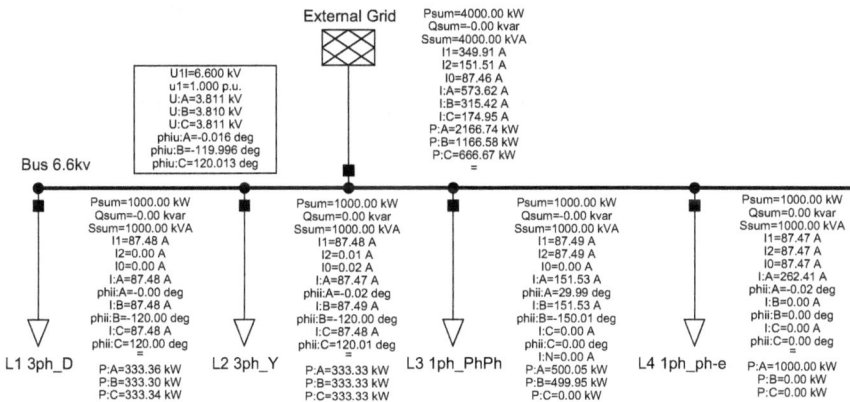

FIGURE 10.2
Four different 1 MW loads.

TABLE 10.1

Characteristics of the Loads in Figures 10.2 and 10.1

Load Title	Specifications
L1 3ph_D	Three-phase symmetrical triangle load
L2 3ph_Y	Three-phase symmetrical star load
L3 1ph_PhPh	Single-phase load between two phases (phases A and B)
L4 1ph_ph-e	Single-phase load between phase A and earth

$$\Rightarrow \hat{I}_A = 87.48\,\text{A}\angle 0,\ \hat{I}_B = 87.48\,\text{A}\angle -120,\ \hat{I}_C = 87.48\,\text{A}\angle 120$$

$$P_A = P_B = P_C = \frac{1\,\text{MW}}{3} = 0.333\ \text{MW}$$

From equation (10.3) we have:

$$
\begin{bmatrix} \hat{I}_a^0 \\ \hat{I}_a^+ \\ \hat{I}_a^- \end{bmatrix} = \frac{1}{3}
\begin{bmatrix} 1 & 1 & 1 \\ 1 & \alpha & \alpha^2 \\ 1 & \alpha^2 & \alpha \end{bmatrix}
\begin{bmatrix} 87.48\angle 0 \\ 87.48\angle -120 \\ 87.48\angle 120 \end{bmatrix} = \frac{87.48}{3}\begin{bmatrix} 0 \\ 3 \\ 0 \end{bmatrix} = \begin{bmatrix} 0 \\ 87.48\text{A}\angle 0 \\ 0 \end{bmatrix}
$$

2. It makes no difference if the load is a star, delta, between two phases, or one phase and ground when the power load is constant (1 MW in this example). It always uses the same amount of power overall.

3. In the load between two phases (L3 1ph_PhPh), the current $(I_A=-I_B)$ is calculated as follows.

$$P = |S| = (V_L)(I) \Rightarrow 1\,\text{MW} = 6.6\,\text{kV}\,(I_A) \Rightarrow I_A = 151.52\,\text{A}$$

$$\hat{V}_A = \frac{6.6\,\text{kV}}{\sqrt{3}}\angle 0 = 3.811,$$

$$\hat{V}_B = 3.811\angle -120 \Rightarrow \hat{V}_{AB} = 6.6\,\text{kV}\angle 30 \Rightarrow \hat{I}_A = -\hat{I}_B = 151.52\angle 30$$

From equation (10.3) we have:

$$
\begin{bmatrix} \hat{I}_a^0 \\ \hat{I}_a^+ \\ \hat{I}_a^- \end{bmatrix} = \frac{1}{3}
\begin{bmatrix} 1 & 1 & 1 \\ 1 & \alpha & \alpha^2 \\ 1 & \alpha^2 & \alpha \end{bmatrix}
\begin{bmatrix} 151.52\angle 30 \\ -151.52\angle 30 \\ 0 \end{bmatrix}
$$

$$
= \frac{151.52\angle 30}{3}\begin{bmatrix} 0 \\ 1-1\angle 120 \\ 1-1\angle 240 \end{bmatrix} = \begin{bmatrix} 0 \\ 87.48\text{A}\angle 0 \\ 87.48\text{A}\angle 60 \end{bmatrix}
$$

4. The single-phase A load calculations are as follows.

$$P = |S| = (V_{ph})(I) \Rightarrow 1\,\text{MW} = 3.811\,\text{kV}\,(I_A) \Rightarrow I_A = 262.43\,\text{A}, \;\; I_B = I_C = 0$$

$$P_A = 1\,\text{MW}, \;\; P_B = P_C = 0$$

From equation (10.3) we have:

$$\begin{bmatrix} \hat{I}_a^0 \\ \hat{I}_a^+ \\ \hat{I}_a^- \end{bmatrix} = \frac{1}{3} \begin{bmatrix} 1 & 1 & 1 \\ 1 & \alpha & \alpha^2 \\ 1 & \alpha^2 & \alpha \end{bmatrix} \begin{bmatrix} 262.43\angle 0 \\ 0 \\ 0 \end{bmatrix} = \frac{262.43\angle 0}{3} \begin{bmatrix} 1 \\ 1 \\ 1 \end{bmatrix} = \begin{bmatrix} 87.48\,\text{A}\angle 0 \\ 87.48\,\text{A}\angle 0 \\ 87.48\,\text{A}\angle 0 \end{bmatrix}$$

5. The law of conservation of power and KCL can be used as follows:

$$P_A = \sum_{i=1}^{4} P_{Ai} = 333.33 + 333.33 + 500 + 1000 = 2166.66\,\text{W}$$

$$P_B = \sum_{i=1}^{4} P_{Bi} = 333.33 + 333.33 + 500 + 0 = 1166.66\,\text{W}$$

$$P_C = \sum_{i=1}^{4} P_{Ci} = 333.33 + 333.33 + 0 + 0 = 666.66\,\text{W}$$

$$\hat{I}_A = \sum_{i=1}^{4} \hat{I}_{Ai} = 87.48\angle 0 + 87.48\angle 0 + 151.52\angle 30 + 262.43\angle 0 = 573.63\,\text{A}\angle 7.59$$

$$\hat{I}_B = \sum_{i=1}^{4} \hat{I}_{Bi} = 87.48\angle -120 + 87.48\angle -120 - 151.52\angle 30 + 0 = 315.41\,\text{A}\angle -133.9$$

$$\hat{I}_C = \sum_{i=1}^{4} \hat{I}_{Ci} = 87.48\angle 120 + 87.48\angle 120 + 0 + 0 = 174.96\,\text{A}\angle 120$$

$$\hat{I}_a^+ = \sum_{i=1}^{4} \hat{I}_{ai}^+ = 87.48\angle 0 + 87.48\angle 0 + 87.48\angle 0 + 87.48\angle 0 = 349.92\,\text{A}$$

$$\hat{I}_a^- = \sum_{i=1}^{4} \hat{I}_{ai}^- = 0 + 0 + 87.48\angle 60 + 87.48\angle 0 = 151.52\,\text{A}\angle 30$$

$$\hat{I}_a^0 = \sum_{i=1}^{4} \hat{I}_{ai}^0 = 0+0+0+87.48\angle 0 = 87.48\,A\angle 0$$

It is evident from the calculations above that calculating the components of electric current under asymmetrical conditions is far easier than calculating asymmetrical electric currents directly. However, the KCL and power relations are valid in every situation. How to use the software and the possibility for further work are explained in the appendix. For more detailed work and further work, you can refer to the references [1–7] at the end of this chapter.

10.4 Two-Choice Questions (Yes/No)

1. Asymmetrical faults are more common than symmetrical faults.
2. Unbalanced loads can cause voltage unbalance in a power system.
3. Symmetrical components are used to analyze balanced power systems only.
4. Positive sequence components represent the balanced component of a three-phase system.
5. Negative sequence components are associated with unbalanced conditions like single-phase faults.
6. Zero sequence components are primarily associated with ground faults.
7. Unbalanced lines can lead to increased power losses in a power system.
8. Power system stabilizers are used to mitigate the effects of unbalanced loads.
9. Symmetrical components can be used to analyze power system transients.
10. Unbalanced loads can cause harmonic distortion in power systems.
11. Zero sequence current can flow in the neutral conductor of a three-phase system.
12. Power system relays can be designed to detect and respond to unbalanced conditions.
13. Unbalanced voltages can lead to increased stress on power system equipment.

14. The negative sequence component of voltage is zero in a balanced three-phase system.

15. The zero-sequence component of current is zero in a balanced three-phase system.

16. Unbalanced loads can reduce the power factor of a power system.

17. Transposition of transmission lines can help to reduce the impact of unbalanced line impedances.

18. Power system stabilizers can be used to improve the damping of power system oscillations caused by unbalanced conditions.

19. Unbalanced loads can lead to increased noise and vibration in power system equipment.

20. The zero-sequence impedance of a transformer is typically much higher than the positive and negative sequence impedances.

21. The negative sequence impedance of a synchronous generator is typically lower than the subtransient positive sequence impedance.

22. Unbalanced voltages can lead to incorrect operation of metering and protection devices.

23. The zero-sequence impedance of a transmission line is primarily inductive.

24. Unbalanced loads can reduce the efficiency of power system operation.

25. The negative sequence component of current can cause additional heating losses in transformers.

10.4.1 Key Answers to Two-Choice Questions

Yes	Others
No	3, 8, 18

10.5 Appendix on Asymmetrical Distribution Networks in DIgSILENT

This section requires a basic understanding of DIgSILENT PowerFactory software. You must download the file (Chapter10.pfd) from the book's end-of-book attachments in order to follow this section. In this section, instead of constructing a network, an asymmetrical network is analyzed. See Figure 10.1 or 10.2. The main 6.6 kV line-to-line bus is connected to an external network. Two loads (L1 3ph_D) and (L2 3ph_Y) are symmetrical delta and star and two loads (L3 1ph_PhPh) and (L4 1ph_ph-e) are asymmetrical. See Table 10.1. Figure 10.3 shows the variations of these four loads.

Make sure to use an unbalanced (three-phase-ABC) load flow to see the results of this season. In normal mode, the information displayed is very limited and you must follow the steps below to display more information.

Step 1.
See Figure 10.4. Right-click on any of the results boxes and check the (Edit Format for...) option (No, 1). After checking (No. 7), Figure 10.5 is shown.

Step 2.
As shown in Figure 10.5, all three-phase unbalanced information can be selected to be displayed on the network.

Step 3.
See Figure 10.6. Then right-click on each of the result boxes and click the Adapt Width button.

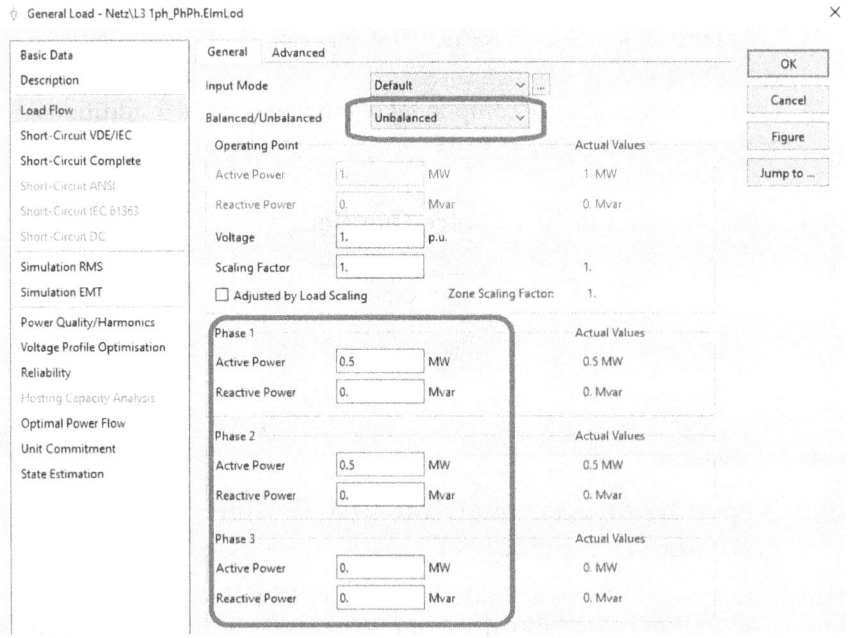

FIGURE 10.3
Symmetrical and asymmetrical load information.

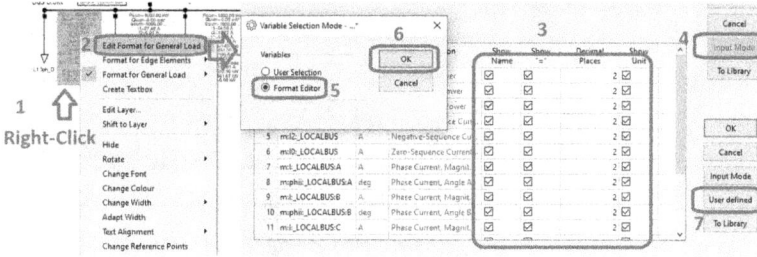

FIGURE 10.4
Steps to add information to the results box, Part 1.

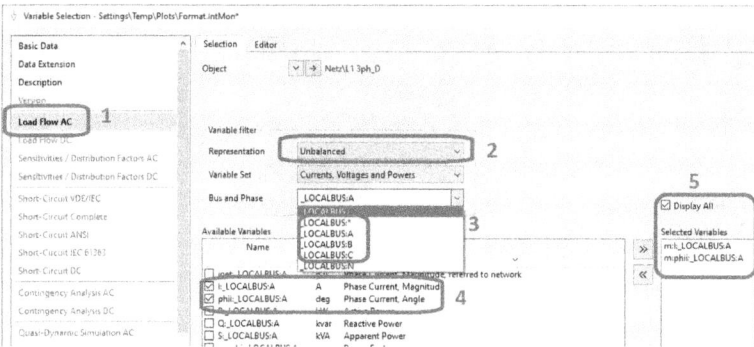

FIGURE 10.5
Steps to add information to the results box, Part 2.

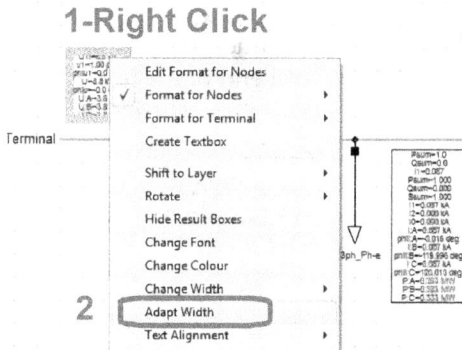

FIGURE 10.6
The **Adapt Width** button.

Step 4.

A three-phase line is modeled in System 1 with the information in Figure 10.7. A single-phase line is modeled in System 2 with the information in Figure 10.8.

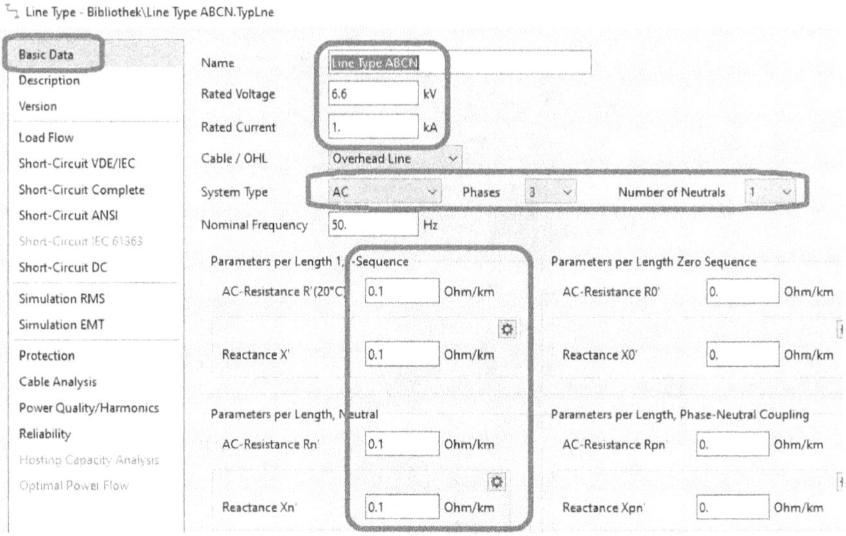
Line Type - Bibliothek\Line Type ABCN.TypLne
Basic Data
Description
Version
Load Flow
Short-Circuit VDE/IEC
Short-Circuit Complete
Short-Circuit ANSI
Short-Circuit IEC 61363
Short-Circuit DC
Simulation RMS
Simulation EMT
Protection
Cable Analysis
Power Quality/Harmonics
Reliability
Hosting Capacity Analysis
Optimal Power Flow
Name
Line Type ABCN
Rated Voltage
6.6
kV
Rated Current
1.
kA
Cable / OHL
Overhead Line
System Type
AC
Phases
3
Number of Neutrals
1
Nominal Frequency
50.
Hz
Parameters per Length 1,2-Sequence
AC-Resistance R'(20°C)
0.1
Ohm/km
Reactance X'
0.1
Ohm/km
Parameters per Length Zero Sequence
AC-Resistance R0'
0.
Ohm/km
Reactance X0'
0.
Ohm/km
Parameters per Length, Neutral
AC-Resistance Rn'
0.1
Ohm/km
Reactance Xn'
0.1
Ohm/km
Parameters per Length, Phase-Neutral Coupling
AC-Resistance Rpn'
0.
Ohm/km
Reactance Xpn'
0.
Ohm/km

FIGURE 10.7
Line type ABCN.

Step 5.

The three-phase load information for System 1 is shown in Figure 10.9, and the three single-phase loads for System 2 are shown in Figure 10.10. Comparing systems 1 and 2 should yield the same result.

Step 6.

Note that in single-phase lines, the phase type (a,b,c) must be known at the beginning and end of the line. Proceed as shown in Figure 10.11.

Step 7.

It is possible to execute an unbalanced load flow in this last step. Systems 1 and 2's responses have to match.

10.6 Summary

Asymmetrical conditions in power distribution networks, primarily due to unbalanced loads and line configurations, pose significant challenges such as increased power losses, voltage imbalances, and potential equipment damage. These imbalances arise from unequal current flow in three-phase systems and can be exacerbated by factors like single-phase

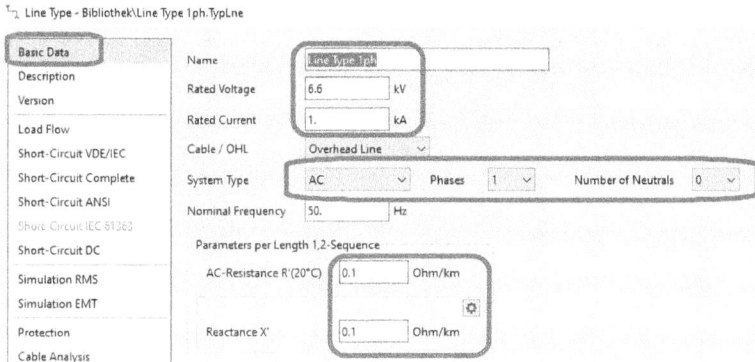

FIGURE 10.8
Line type 1-phase.

FIGURE 10.9
The three-phase load information for System 1.

FIGURE 10.10
Information on three single-phase loads of system 2.

loads and faulty connections. To address these issues, engineers can employ techniques such as load balancing and the use of specialized transformers. This chapter outlines the analysis of these unbalanced networks, their impacts, and provides guidance on simulating them using DIgSILENT software, with references for further exploration.

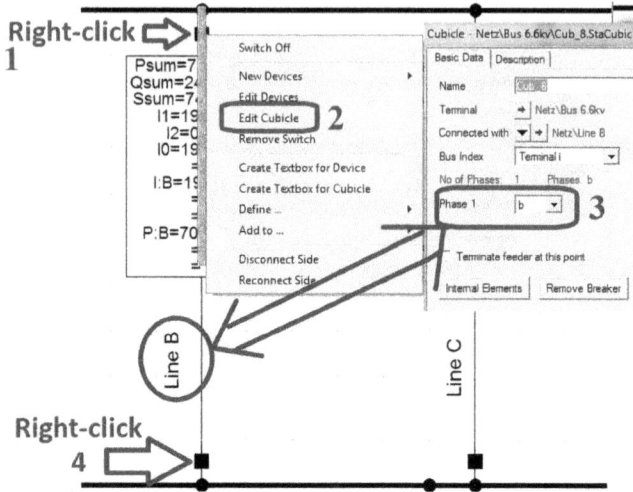

FIGURE 10.11
Determining the type of phase in single-phase lines.

Note

1 The method of *symmetrical components* was introduced by **Charles L. Fortescue** in 1918 in his seminal paper, *"Method of Symmetrical Co-ordinates Applied to the Solution of Polyphase Networks."* This technique revolutionized the analysis of unbalanced polyphase power systems.

References

1. Mokryani, G. *Future Distribution Networks: Planning, Operation, and Control,* AIP Publishing LLC, 2022, https://doi.org/10.1063/9780735422339
2. Blackburn, J.L. *Symmetrical Components for Power Systems Engineering,* CRC Press, 2017.
3. Eidiani, M., "A reliable and efficient method for assessing voltage stability in transmission and distribution networks," *International Journal of Electrical Power and Energy Systems,* 2011, 33(3), pp. 453–456, https://doi.org/10.1016/j.ijepes.2010.10.007
4. Eidiani, M., "A new method for assessment of voltage stability in transmission and distribution networks," *International Review of Electrical Engineering,* 2010, 5(1), pp. 234–240.

5. Eidiani, M., Rouzbehi, K. *Fundamentals of Power System Transformers Modeling, Analysis, and Operation,* Taylor & Francis Group, CRC Press, 2025, pp. 1–127. http://www.routledge.com/9781032881751

6. Parhamfar, M., Eidiani, M., Abtahi, M. "Distributed energy storage system: Case study," In: Hossain, E., Ghosh, A. (eds) *Distributed Energy Storage Systems for Digital Power Systems.* Elsevier, 2024, pp. 395–422. https://doi.org/10.1016/B978-0-443-22013-5.00013-7

7. Eidiani, M., Zeynal, H., "Flexible interconnection in energy systems via variable frequency transformer," *Majlesi Journal of Energy Management,* 8(3), pp. 45–53, 2019.

11

Optimal Power Flow Analysis

11.1 Introduction

One of the most well-known forms of optimal power flow (optimal load flow) to reduce costs or increase profits, lower overall network losses, etc., is Economic (optimal) Load Dispatch. Power plant cost functions are typically modeled using a quadratic equation, and optimization techniques like Lagrange and Kuhn–Tucker are used to solve this problem.

The optimal power flow (OPF) problem is solved by taking into account a number of constraints. The OPF problem can be solved by taking into account the constraints on the generation of active and reactive power by generators, the transmission power constraints of lines, transformer power, and bus voltage.

Various methods have been developed to solve the OPF problem, including the Newton-Raphson method, interior point method, linear programming, and nonlinear programming. However, the OPF problem is a challenging problem due to its nonlinear nature, large scale, multiple objectives, dynamic constraints, and uncertainty. To address these challenges, researchers have explored advanced techniques like metaheuristic algorithms, machine learning, robust optimization, and distributed optimization.

The OPF problem can be formulated as a nonlinear optimization problem with an objective function, equality constraints, and inequality constraints. By solving this problem, we can determine the optimal settings for control variables to achieve the desired objective while satisfying all operational limits [1–11].

In this section, OPF in DIgSILENT PowerFactory is examined in all of its specifics and constraints. Working with DIgSILENT is taught, and the OPF concept is reviewed. How to use the software and the possibility of doing further work are explained in the appendix. For more detailed work and further work, you can refer to the references [1–11] at the end of this chapter and the software user manual.

DOI: 10.1201/9781003590514-11

11.2 Optimal Power Flow

This section focuses on details of the optimal power flow in the DIgSILENT program. Refer to the **Optimal Power flow/Unit Commitment** of the software in Figure 11.1. Following numbers 1 and 2 in Figure 11.1, the OPF window will appear. After entering the OPF window (No. 4), in the initial **Basic Options** section (No. 5), OPF methods are first seen. The AC optimization method, which is the most accurate method, is usually used [1–11].

The DC optimization approach can be applied in situations where the network is large and problem-solving speed is crucial, like in **Unit Commitment**. The software can minimize or maximize seven objective functions, such as cost, losses, reactive power, etc., as shown in No. 7 in the figure. The most common objective function is typically to minimize the overall cost.

Additionally (No. 8), the software has the option to consider or ignore controls like tap transformers, shunt reactors, capacitors, and the active and reactive power of generators and SVCs. Lastly (No. 9), the software has the option to take into account or ignore constraints such as the active and reactive power of SVCs and generators, all element load limits, bus voltage limits, and activated boundary flow limits.

Next, the information type of each element in the network is examined.

> **Loads:** The permissible voltage range in the OPF needs to be specified in this element, as shown in Figure 11.2. Both upper and lower bounds can be specified, and this restriction can be added to or removed from OPF.

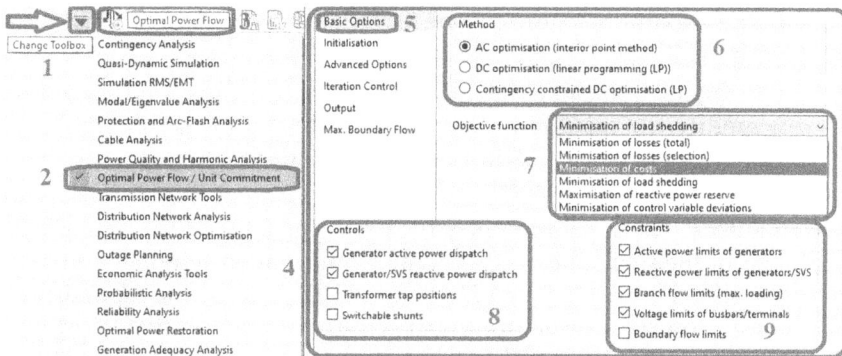

FIGURE 11.1
The OPF window, methods, controls, and constraints.

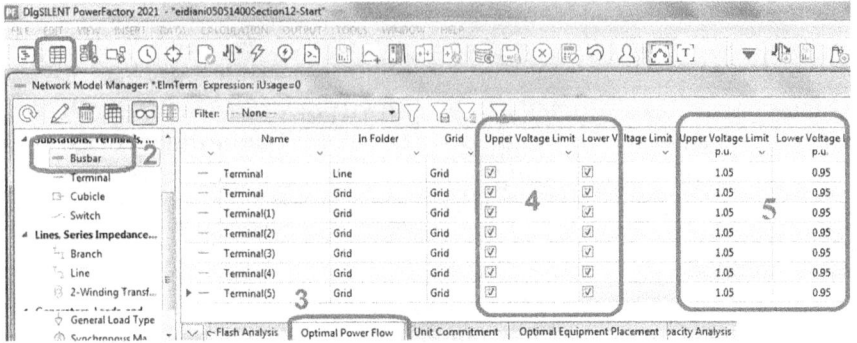

FIGURE 11.2
Upper and lower voltages limits window.

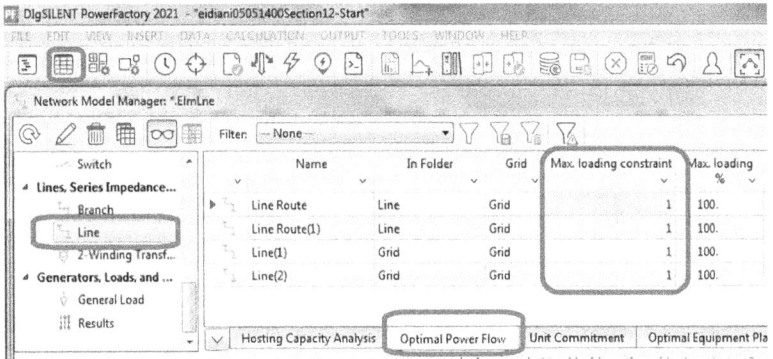

FIGURE 11.3
Maximum line loading constraint window.

Lines: You can either add or remove this constraint in OPF or modify the maximum loading from 100% in this element, as shown in Figure 11.3.

Transformers: You can either add or remove this constraint in OPF or modify the maximum loading from 100% in this element, as shown in Figure 11.4. Additionally, the transformer tap position can be enabled or disabled.

Generators: The synchronous generator is the most crucial component of the OPF. The OPF generator information window is displayed in Figures 11.5 and 11.6. Figure 11.5 displays the capability curve, reserve power optimization, active and reactive power controls, and active and reactive power operational limits. The type of analysis and investigation of the OPF problem determines the values of the controls and constraints in Figure 11.5.

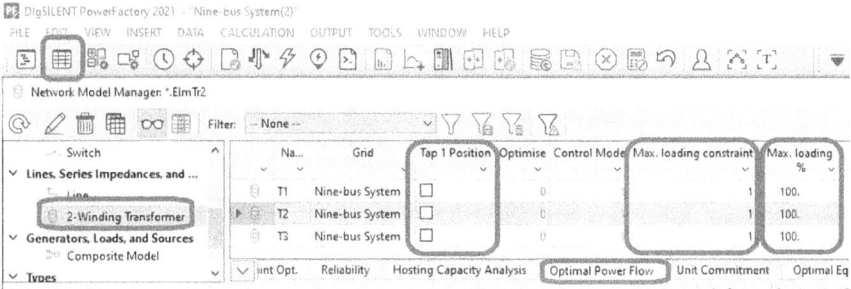

FIGURE 11.4
Maximum transformer loading constraint window.

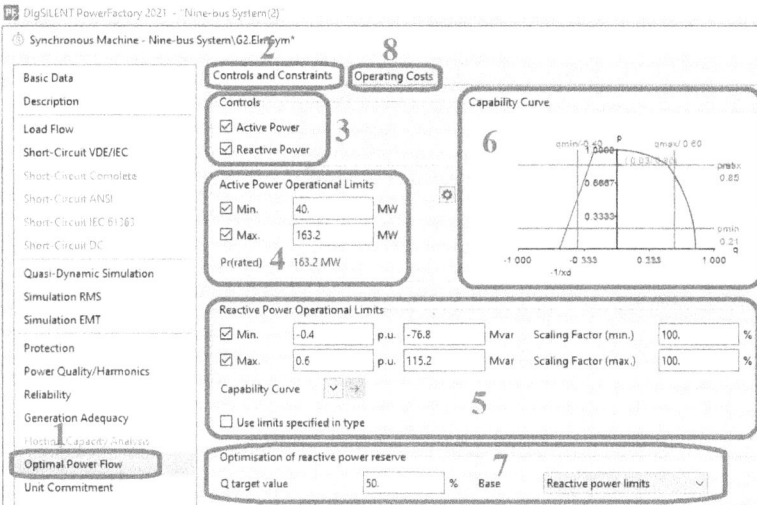

FIGURE 11.5
Controls and constraints of generator window.

Refer to Figure 11.6. There are four ways to estimate the generator operating cost in this window. No. 2 in this figure indicates that the Polynomial degree 2 function is the most widely used approximation of the cost function. More rows can be added to No. 3 to increase the number of operating points. This curve can be seen in No. 4 by calculating the type of cost function approximation and adding the number of operating points. Lastly, the values of fixed and penalty costs can be found in No. 5.

Typically, industrial software like DIgSILENT provides basic data for various OPF element values. Thus, information that is unknown in a section of the network can also be utilized.

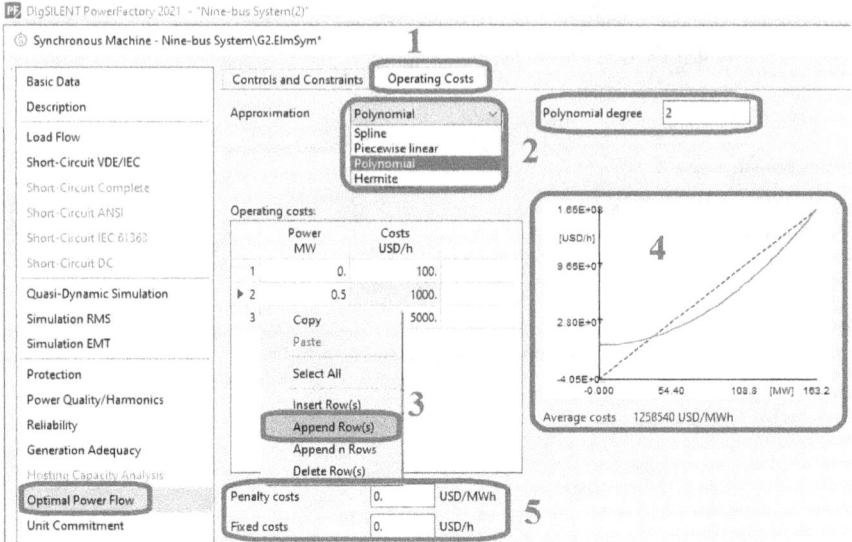

FIGURE 11.6
Operation costs of generator window.

11.3 Two-Choice Questions (Yes/No)

1. The primary goal of OPF is to minimize power system losses.
2. OPF is a static optimization problem.
3. The Newton-Raphson method is a popular technique for solving OPF problems.
4. OPF can be used to improve the voltage profile in a power system.
5. The objective function in OPF can be a combination of multiple factors, such as generation cost and transmission losses.
6. OPF can be used to determine the optimal settings of control variables like generator outputs and transformer tap positions.
7. The power flow equations are linear, making the OPF problem straightforward to solve.
8. OPF can be used to analyze the impact of renewable energy integration on power system operation.
9. The interior point method is a commonly used technique for solving large-scale OPF problems.
10. OPF solutions are always optimal, regardless of the complexity of the power system.

11. Uncertainty in load demand and generation can impact the accuracy of OPF solutions.

12. OPF can be used to improve the stability of a power system.

13. The objective function in OPF is always expressed as a minimization problem.

14. The power flow equations are nonlinear, making the OPF problem challenging to solve.

15. Metaheuristic algorithms can be used to solve large-scale OPF problems with complex constraints.

16. The optimal solution to an OPF problem is always unique.

17. OPF can be used to identify potential congestion points in a power system.

18. The power flow equations are used to model the steady-state behavior of a power system.

19. OPF can be used to determine the optimal location for new power plants.

20. The objective function of OPF can include environmental factors, such as emissions.

21. Real-time OPF can be used to adapt to changing system conditions.

22. The Newton-Raphson method is guaranteed to converge to the global optimum.

23. The interior point method is more efficient than the Newton-Raphson method for large-scale OPF problems.

24. OPF can be used to coordinate the operation of multiple power systems.

25. The optimal solution to an OPF problem may not be feasible due to practical constraints.

26. The power flow equations are based on Kirchhoff's laws.

27. OPF can be used to analyze the impact of different operating scenarios on power system performance.

28. The objective function of OPF can include economic factors, such as fuel costs.

29. The power flow equations are used to calculate the voltage magnitudes and phase angles at each bus in a power system.

30. OPF can be used to identify potential voltage stability problems.

31. The objective function of OPF can include social factors, such as environmental impact.

32. The power flow equations are derived from the application of Kirchhoff's voltage and current laws to the power system network.

33. OPF can be used to determine the optimal reactive power compensation scheme for a power system.

34. The interior point method is a type of linear programming technique.

35. The Newton-Raphson method can be used to solve both AC and DC power flow problems.

36. OPF can be used to analyze the impact of distributed generation on power system operation.

37. The objective function of OPF can include the cost of energy storage.

38. The power flow equations can be used to model the dynamic behavior of a power system.

39. OPF can be used to determine the optimal location and size of new transmission lines.

40. The interior point method is more robust to initial conditions than the Newton-Raphson method.

11.3.1   Key Answers to Two-Choice Questions

Yes	Others
No	1, 7, 10, 13, 16, 22, 34, 38

11.4 Appendix, OPF in DIgSILENT

A basic understanding of DIgSILENT PowerFactory software is required for this section. To follow this section, download the file (Chapter11.pfd) from the book's end-of-book attachments. Examine Figure 11.7 after running the load flow. Please review Figure 11.7's graphical network output data. The network losses are now 2.46 MW in Figure 11.8.

Turn on the transformers' maximum loading constraint (Figure 11.4), the lines' maximum loading constraint (Figure 11.3), and the bus's upper and lower voltage limits (Figure 11.2). The cost function and the production units' limitations can be specified in this section. Following Figure 11.9, double-click each generator to open the OPF window and enter the appropriate values. Please take note that Figure 11.9 is presented on two different pages starting with the 2016 editions.

The cost function variable is added to the list of variables that generators can display in this section. After running the load flow, proceed as outlined in Figure 11.10 through number 10. The figure should display the cost of generation in "base case" mode.

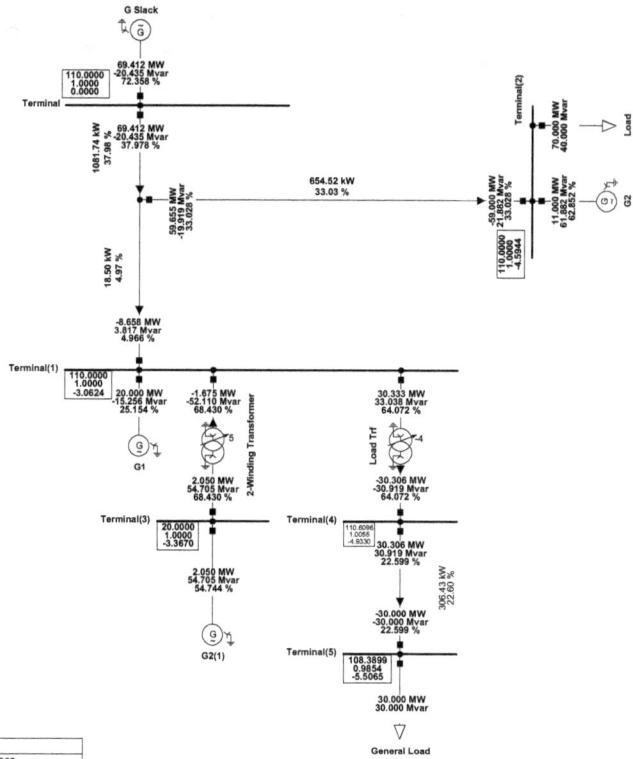

FIGURE 11.7
Typical load flow of the OPF network case study.

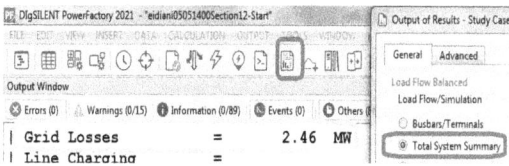

FIGURE 11.8
Grid losses in **total system summary**.

If you use the mouse to select all four generators' production costs in No. 10 in Figure 11.10, information No. 11 will appear. It determines the generator's total, average, minimum, and maximum costs. The total cost of production is $53.649 per hour. At last, we reach the last section of the work. As illustrated in Figure 11.1, run the OPF. Indicate the optimization in terms of the lowest possible cost. As illustrated in Figure 11.11, you can

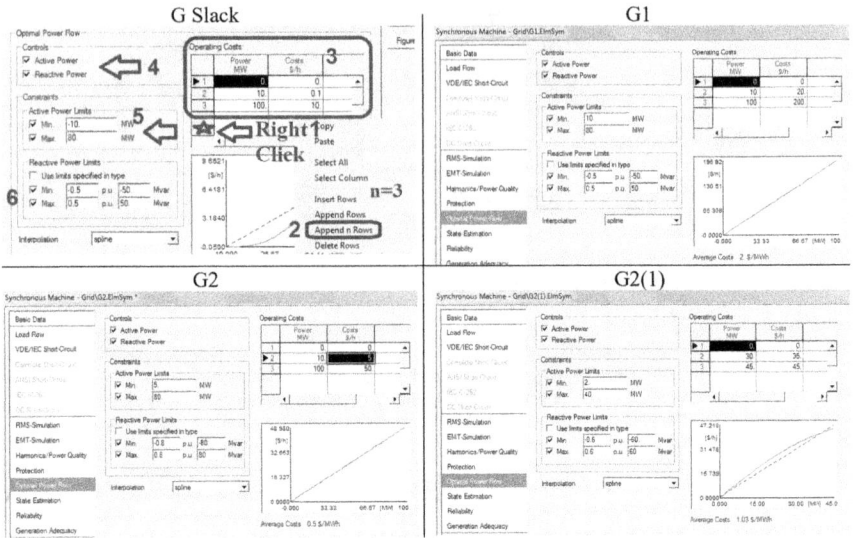

FIGURE 11.9
Operation costs of four generators in study case.

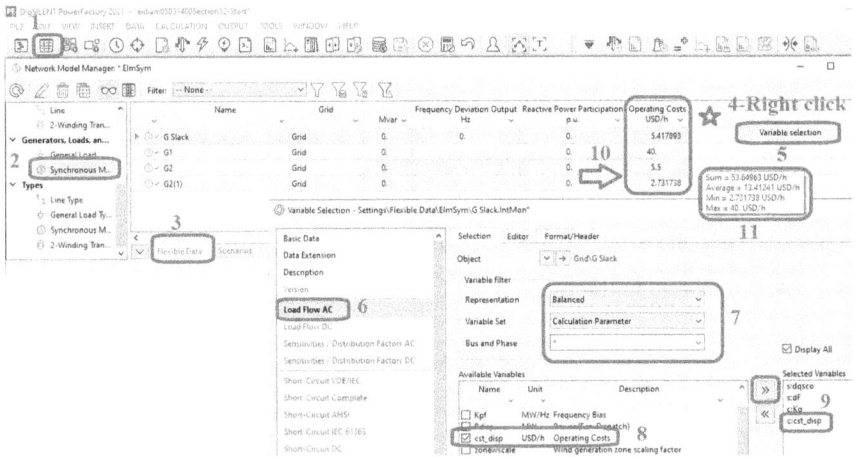

FIGURE 11.10
Enable operation costs in load flow in study case.

modify "Initialization of AC -optimization" if you observe a convergence error.

Return to the "Flexible Data" section (No. 3 in Figure 11.10) to view the power generators' optimal cost values, which are displayed in Figure 11.12. In Figure 10.11, repeat steps 4 and beyond if the cost values are not displayed. Compared to the base case cost, the total cost of generation after

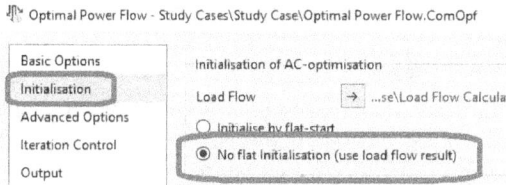

FIGURE 11.11
Initialization of AC -optimization.

FIGURE 11.12
OPF output.

OPF is **34.8288 \$/h**. The network losses are now displayed as **2.38 MW** in Figure 11.8. Although losses have decreased, it should be noted that the objective was to lower overall costs rather than losses.

Reset the objective function to OPF to "minimization of losses" for the last analysis. The final response includes the losses (0.33 MW) and the total cost (\$93.04/h). Even though the losses have been decreased, the solution is unrealistic and unreasonable.

11.5 Summary

Optimal Power Flow (OPF) is a crucial concept in power systems aimed at minimizing costs and losses while adhering to various operational constraints, such as power generation limits and transmission capabilities. It employs quadratic cost functions and optimization techniques like Lagrange multipliers and Kuhn–Tucker conditions, but faces challenges due to its nonlinear nature and multiple objectives. Various methods, including Newton-Raphson and machine learning, have been developed to tackle these complexities. The text also discusses the application of OPF within the DIgSILENT PowerFactory software, providing guidance on its use and further exploration of the topic.

References

1. Zeynal, H., Jiazhen, Y., Azzopardi, B., Eidiani, M. "Flexible economic load dispatch integrating electric vehicles," *2014 IEEE 8th International Power Engineering and Optimization Conference (PEOCO2014)*, 2014, pp. 520–525, https://doi.org/10.1109/PEOCO.2014.6814484

2. Zeynal, H., Hui, L.X., Jiazhen, Y., Eidiani, M., Azzopardi, B. "Improving Lagrangian Relaxation Unit commitment with Cuckoo Search Algorithm," *2014 IEEE International Conference on Power and Energy (PECon)*, 2014, pp. 77–82, https://doi.org/10.1109/PECON.2014.7062417

3. Zeynal, H., Eidiani, M. "Hydrothermal scheduling flexibility enhancement with pumped-storage units," *2014 22nd Iranian Conference on Electrical Engineering (ICEE)*, 2014, pp. 820–825, https://doi.org/10.1109/IranianCEE.2014.6999649

4. Eidiani, M., Heidari, V., *Fundamentals of Power Systems Analysis 1: Problems and Solutions*, Taylor & Francis Group, CRC Press, 2023, pp. 1–215, https://doi.org/10.1201/9781003394433

5. Eidiani, M., Rouzbehi, K., *Advanced Topics in Power Systems Analysis: Problems, Methods, and Solutions*, Taylor & Francis Group, CRC Press, 2024, pp. 1–120. http://www.routledge.com/9781032828664

6. Eidiani, M., "An efficient differential equation load flow method to assess dynamic available transfer capability with wind farms," *IET Renewable Power Generation*, 2021, 17(5), pp. 3843–3855, https://doi.org/10.1049/rpg2.12299

7. Eidiani, M., "A new load flow method to assess the static available transfer capability," *Journal of Electrical Engineering and Technology*, 2022, 17(5), pp. 2693–2701, https://doi.org/10.1007/s42835-022-01105-3

8. Raza, A., Mustafa, A., Rouzbehi, K., Jamil, M., Gilani, S. O., Abbas, G., Farooq, U., "Optimal power flow and unified control strategy for multi-terminal HVDC systems," *IEEE Access*, vol. 7, pp. 92642–92650, 2019.

9. Pourmirasghariyan, M., Milimonfared, J., Yazdi, S. S. H., Rouzbehi, K., "Hybrid method for optimal placement of DC-PFCs to enhance static security of VSC-HVdc grids," *IEEE Systems Journal*, vol. 16, no. 3, pp. 4839–4848, 2021.

10. Pourmirasghariyan, M., Milimonfared, J., Yazdi, S. S. H., Biglo, A. H. A., Rouzbehi, K., "Application of max flow-min cut theory to find the best placement of electronic-based DC-PFCs for enhancing static security in MT-HVDC meshed grids," *2022 30th International Conference on Electrical Engineering (ICEE)*, pp. 23–28, 2022.

11. Rouzbehi, K., Baa Wafaa, M., Rakhshani, E., "An imperialist competitive algorithm-based multi-objective optimization for voltage source converter high-voltage direct current stations control in multi-terminal HVDC grids," *Electric Power Components and Systems*, vol. 47, no. 4-5, pp. 316–328, 2019.

12

Reliability and Contingency Analysis

12.1 Introduction

This chapter demonstrates reliability analysis techniques using DIgSILENT PowerFactory software. It provides essential concepts, methodologies, and practical guidance for conducting comprehensive reliability assessments of power systems. Reliability stands as a cornerstone in power system planning and operations, fundamentally shaping decisions about system design, maintenance, and real-time control.

System reliability quantifies the probability that a component or system will satisfactorily perform its intended functions within specified parameters over a defined period under stated operating conditions. The four primary components of this definition are time, specified operating conditions, probability, and satisfactory performance.

The four core components that define reliability are:

1. Probability: The mathematical likelihood of system operation
2. Time: The specified duration over which reliability is measured
3. Operating Conditions: The defined environmental and load parameters under which the system must function
4. Satisfactory Performance: The criteria that determine acceptable system operation.

The term "reliability" encompasses a wide range of meanings and requires further subdivision to enhance clarity and comprehension. A practical approach is to break it into two key concepts: **System Security** and **System Adequacy**. This straightforward yet effective division is elaborated upon in the following section.

After that, you will study two crucial and useful ideas: **Energy Not Supplied (ENS)** and **Contingency Analysis**.

How to use the software and the possibility of doing further work are explained in the appendix. For more detailed work and further work, you can refer to the references [1–7] at the end of this chapter.

DOI: 10.1201/9781003590514-12

12.2 Reliability Definition

Two essential ideas for planners and operators of power systems are adequacy and security. The existence of enough equipment in the system to satisfy the degree of customer demand is correlated with system adequacy. Having enough generating units and the required transmission and distribution networks to move electrical energy from the producer to the consumer is a component of system adequacy. Stated differently, system adequacy is associated with the system's steady-state conditions.

However, system security assessment evaluates how well the system can handle the emergence and growth of power system disruptions and disturbances. These include circumstances pertaining to regional disruptions or worldwide, all-encompassing disruptions, and outages in transmission line equipment and generation. We employ various computational techniques to evaluate system security and system adequacy. System adequacy assessment includes, for instance, determining the expected energy outages (Loss of Energy Expectation) (LOEE) and the number of outages in the desired period (Loss of Load Expectation) (LOLE).

12.2.1 Reliability Assessment Classification

In contemporary systems, power transmission from power plant generating units to individual consumers is so vast and complex that it is not feasible to provide an accurate and comprehensive assessment of the reliability of this massive network with the hardware and software currently in use. To address this issue, the power system network's components are separated into three categories based on their respective functions: generation, transmission, and distribution. By combining these functional domains, hierarchical levels for power system reliability analysis can be produced. Figure 12.1 shows the functional domains and hierarchical levels of reliability assessment.

Only the generation unit level of reliability assessment is covered by the first hierarchical level (HLI). This level of reliability assessment entails a quantitative analysis of the generating units' capacity to supply the power system's overall load. The transmission line equipment is included in the

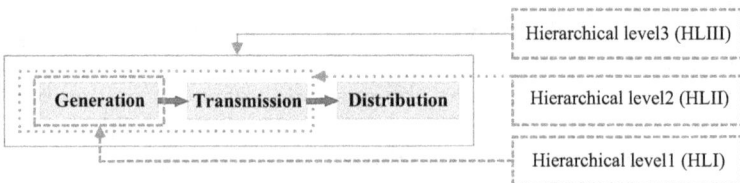

FIGURE 12.1
Hierarchical levels in reliability assessment.

second hierarchical level (HLII) reliability assessment in addition to the generating units. A composite system is the result of combining transmission and generation. The ability of the generation and transmission system to simultaneously produce and deliver electrical energy to consumers or the main load points is evaluated when evaluating the reliability of composite systems. The third hierarchical level (HLIII) reliability assessment evaluates the adequacy at the consumer load point level by conducting a reliability study of the entire power system network, or three performance areas. Practically speaking, the reliability indices determined at the three levels differ from one to another. Typically, one or more indices that outline the system's expected performance are used to predict system reliability, and the acceptable values of one of the indices are used as a basis.

You can refer to references [1–6] to gain a better understanding of various reliability concepts. The following demonstrates how to learn how to apply reliability in a network without having to understand additional concepts and with the aid of DIgSILENT software.

12.3 Contingency Analysis

Contingency Analysis in power systems is a predictive study used to evaluate the security and reliability of the electrical grid under possible failure scenarios. In simple terms, it answers the question: "If something in the network fails, will the system still operate safely and within limits?" It is assumed in contingency analysis that every component of the circuit can be taken out. The load flow identifies areas of the network with overload or voltage drops and increases after each element is eliminated. The significance of these issues is used to rank them. The fact that contingency analysis is repeated for every load point and year-round demonstrates how crucial it is to operation.

In power systems, contingency analysis is essential because it evaluates the system's resilience for possible disruptions, such as equipment failures or abrupt changes in load. To ensure the dependability and security of the power grid, engineers can simulate these situations to find weaknesses, assess the effects of outages, and create risk-reduction plans.

It is referred to as $(n-k)$ in contingency analysis if "k" elements are eliminated from the network at any point. The majority of actual networks are designed and operated for $(n-1)$ (Figure 12.2). The Outage level (No. 2) in Figure 12.2 is easily changeable in the DIgSILENT software. Additionally, you can indicate the kind of element being removed (No. 3).

Assume that the contingency ranking is carried out for a nine-bus network that has three transformers, six lines, and three generators. In this example, twelve contingencies can be taken into account. Every one of these

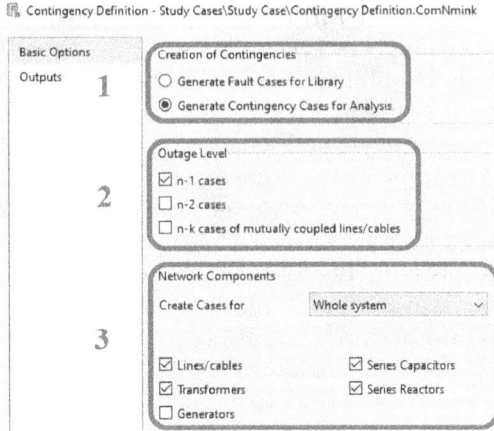

FIGURE 12.2
Basic information on contingency ranking.

scenarios is automatically taken into account by the software. By default, the following items are included in the software, which can be changed.

Record thermal loadings above 80%, Record voltages below 0.95 pu, Record voltages above 1.05 pu, and Record voltage step changes above 5% (Figure 12.3). The output types after contingency ranking calculations can be specified in the software. In this section, the outputs of Voltage Steps, Non-convergent, and loading violations cases are examined. Figure 12.4 shows the outputs of Voltage Steps. In this figure, we have:

1. In the Component column, the bus or terminal that has encountered a problem is sorted and seen in order of critical status.
2. In the Voltage Step column, the difference between columns 3 and 4 is given.

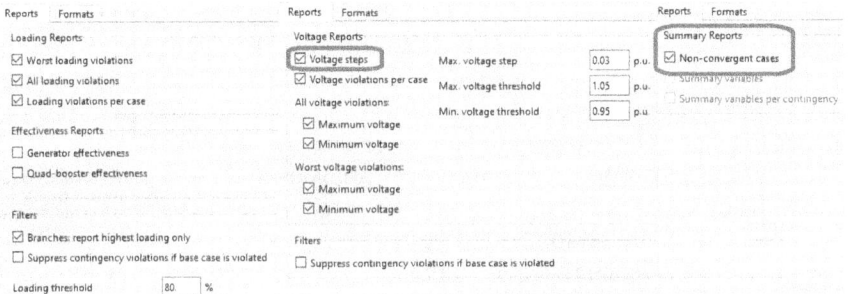

FIGURE 12.3
The output types after contingency ranking.

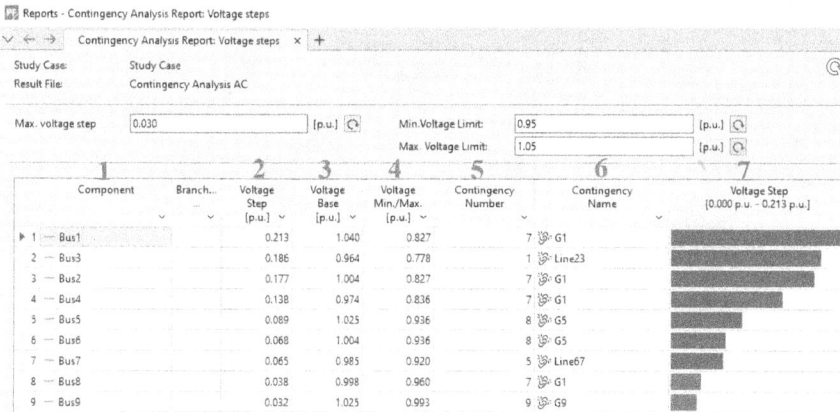

FIGURE 12.4
The voltage steps output.

3. In the Voltage Base column, the bus voltage in the base case is given before a fault occurs in the network.

4. In the Voltage Min./Max column, the bus voltage after the element in column 6 is removed is shown.

5. In the Contingency Number column, the contingency number is given.

6. In the Contingency Name column, the element of the network that has been removed from the circuit is given.

7. In the last column, the same second column is drawn graphically.

The loss of generator G1 is the worst fault in terms of voltage, as shown in Figure 12.4. A 0.213 pu voltage drop results from this fault. In the same way, the following errors are listed in the table.

Figure 12.5 shows the outputs of non-convergent cases. This output shows that the network will diverge (be non-convergent) and exit synchronous mode if certain components are removed from the circuit. It is evident that the network will diverge if two transformers, T56 and T98, are removed from the circuit. Because of this, most networks use two parallel transformers, which allow the other to be replaced if one is removed from the circuit.

The results of the loading violations are displayed in Figure 12.6. It indicates which other elements are having over-loading issues for each disconnected element. As can be observed, the T56 loading increases from 71% to 134.8% -the worst contingency- when the transformer T12 is disconnected.

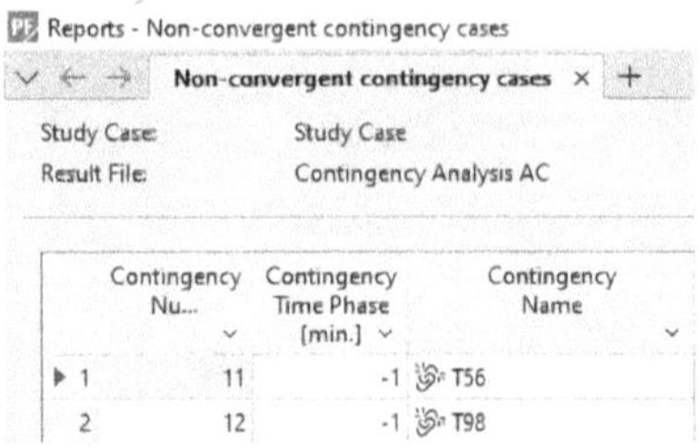

FIGURE 12.5
The non-convergent cases output.

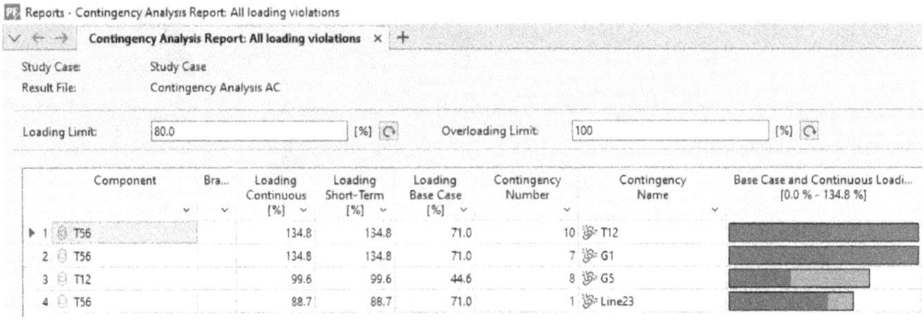

FIGURE 12.6
The loading violations output.

12.3.1 (n-2) Case in Contingency Analysis

The number of contingencies for 12 elements is as follows if we wish to perform $(n-2)$ contingency with $(n-1)$.

$$(n-1) \text{ contingencies}: \binom{12}{1} = \frac{12!}{1!(12-1)!} = 12$$

$$(n-2) \text{ contingencies}: \binom{12}{2} = \frac{12!}{2!(12-2)!} = \frac{12!}{2(10!)} = \frac{(12)(11)}{2} = 66$$

Total contingencies: $12+66=78$

Two components are eliminated from the network simultaneously in case $(n-2)$, which is far less likely and increases the likelihood of a network failure.

12.4 Reliability Analysis

The data needed for reliability calculations for each element is first displayed in this section (Table 12.1). The statistical average of failures over the previous years serves as the basis for these values.

TABLE 12.1

The Data Needed for Reliability Calculations

Terminal or Bus	Failure Frequency for Terminal (1/a)	0.05	Annually
	Additional failure frequency per connection (1/a)	0.05	Annually
	Repair Duration (h)	10	Hour
Line	Failure Frequency (1/a)	0.01	Annually
	Repair Duration (h)	10	Hour
Transformer	Failure Frequency (1/a)	0.01	Annually
	Repair Duration (h)	10	Hour

There are two kinds of failure frequencies for every terminal or bus: the bus failure frequency and the connection failure frequency. Keep in mind that the total of these two failure frequencies represents the busbar failure probability. A busbar with two connections, for instance, with a failure frequency of 0.005 for the busbar and 0.003 for each connection, has a total failure probability of $0.005 + 2 \times 0.003 = 0.011$.

Please take note that we omitted data about loads and generators. This information can be found in **Generation Adequacy**, another type of reliability.

The software can be used to determine system reliability based on information about transformer, load, and line faults. An example of System Summary reliability data is shown in Table 12.2. We pay particular attention to Energy Not Supplied (ENS) among the different output data and indices.

ENS indicates the amount of energy that the network has not been able to sell in a year, which in this case is 4882.5 MWh annually, taking into account the potential for failure of different network components. You can quickly grasp this by looking at the Load Point Energy Not Supplied (LPENS) variable and the Load Interruptions Table 12.3. In fact, LPENS is the ENS calculated separately for each load. Note that the sum of LPENS is the same as the ENS of the entire network. We have:

$$LPENS_{Load3} + LPENS_{Load4} + LPENS_{Load7} = ENS$$

$$1937.50 + 1395.00 + 1550.00 = 4882.5$$

The ENS caused by lines, loads, and transformers can be computed independently in a different analysis. Table 12.4 displays the results of these computations. The ENS of the entire network is derived from the sum of the ENS of each element, as can be seen in this table.

The ENS derived from the line analysis is zero, as shown in Table 12.4. This implies that energy can be sold to the loads and that no load is disconnected for every line loss. Thus, there isn't any energy that isn't distributed. As an exercise, you can now disconnect one of the lines by opening

TABLE 12.2

System Summary Reliability Data

System Summary			
System Average Interruption Frequency Index	:	SAIFI =	1.550000 1/Ca
Customer Average Interruption Frequency Index	:	CAIFI =	1.550000 1/Ca
System Average Interruption Duration Index	:	SAIDI =	15.500 h/Ca
Customer Average Interruption Duration Index	:	CAIDI =	10.000 h
Average Service Availability Index	:	ASAI =	0.9982305936
Average Service Unavailability Index	:	ASUI =	0.0017694064
Energy Not Supplied	:	**ENS =**	**4882.500 MWh/a**
Average Energy Not Supplied	:	AENS =	1627.500 MWh/Ca
Average Customer Curtailment Index	:	ACCI =	1692.397 MWh/Ca
Expected Interruption Cost	:	EIC =	0.000 M$/a
Interrupted Energy Assessment Rate	:	IEAR =	0.000 $/kWh
System energy shed	:	SES =	0.000 MWh/a
Average System Interruption Frequency Index	:	ASIFI =	1.550000 1/a
Average System Interruption Duration Index	:	ASIDI =	15.499999 h/a
Momentary Average Interruption Frequency Index	:	MAIFI =	0.000000 1/Ca

TABLE 12.3

The Load Interruptions Output

						Annex:	Step 1	/ 1
Load Interruptions		TCIT	TCIF	AID	LPENS	LPEIC	ACIF	ACIT
Name		Ch/a	C/a	h	MWh/a	$/a	1/a	h/a
Load3		15.50	1.55	10.00	1937.50	0.00	1.55	15.50
Load4		15.50	1.55	10.00	1395.00	0.00	1.55	15.50
Load7		15.50	1.55	10.00	1550.00	0.00	1.55	15.50

Study Case: Study Case

TABLE 12.4

Calculating ENS Separately for Each Element

ENS	Element
ENS=1732.500 MWh/a	☑ Busbars / terminals ☐ Lines / cables ☐ Transformers
ENS=0.000 MWh/a	☐ Busbars / terminals ☑ Lines / cables ☐ Transformers
ENS=3150.000 MWh/a	☐ Busbars / terminals ☐ Lines / cables ☑ Transformers
The sum of ENS's=4882.5	☑ Busbars / terminals ☑ Lines / cables ☑ Transformers

the switches at both ends. First, the ENS caused by the lines does not go to zero in this instance, and second, the total ENS of the network is equal to the sum of the ENSs.

12.5 Conclusion

In summary, calculating the network's ENS under various conditions is sufficient to assess the network's reliability. It is as simple as that: the lower the ENS value, the more reliable the network is.

12.6 Two-Choice Questions (Yes/No)

1. Power system reliability is the ability of a power system to deliver electricity to consumers with an acceptable level of interruptions.
2. Frequency stability is not a major concern in power system reliability.
3. Voltage stability is a major concern in power system reliability.
4. A power system with higher reliability always has lower costs.

5. Load forecasting is crucial for maintaining power system reliability.

6. Reactive power does not play a significant role in power system reliability.

7. Contingency analysis is a technique used to assess the impact of potential disturbances on power system reliability.

8. Overvoltage is not a potential threat to power system reliability.

9. Under-voltage can lead to equipment damage and power outages.

10. A power system with a high level of redundancy is less reliable.

11. Power system reliability can be improved through preventive maintenance.

12. Aging infrastructure does not impact power system reliability.

13. Extreme weather events can significantly impact power system reliability.

14. Cyberattacks do not pose a threat to power system reliability.

15. Power system operators use real-time monitoring to maintain reliability.

16. A single point of failure can never cause a widespread blackout.

17. Distributed generation can improve power system reliability.

18. Demand-side management does not contribute to power system reliability.

19. Power system reliability standards are essential for maintaining system integrity.

20. The cost of power outages is solely borne by utility companies.

21. Power system reliability is a complex issue with multiple factors influencing it.

22. The reliability of a power system is solely determined by the reliability of its components.

23. The reliability of a power system can be improved through the use of advanced technologies.

24. The reliability of a power system is not affected by human error.

25. Power system reliability is a global concern.

26. The reliability of a power system is not affected by economic factors.

27. The reliability of a power system can be improved through effective communication and coordination.

28. The reliability of a power system is not affected by regulatory policies.

29. The reliability of a power system is influenced by the geographic location of the system.
30. The reliability of a power system is not affected by the type of generation sources used.
31. The reliability of a power system is affected by the age of the infrastructure.
32. The reliability of a power system is not affected by the level of maintenance performed on the system.
33. The reliability of a power system is affected by the level of security measures implemented.
34. The reliability of a power system is not affected by the level of customer awareness and preparedness.
35. The reliability of a power system is a critical factor in the overall economic health of a region.

12.6.1 Key Answers to Two-Choice Questions

Yes	Odd
No	Even

12.7 Appendix, Reliability Analysis in DIgSILENT

This section requires a basic understanding of DIgSILENT PowerFactory software. You must download the file (Chapter8.pfd) from the book's end-of-book attachments in order to follow this section.

Step 1:
Import the file (Chapter8.pfd) and run the load flow and compare the network outputs in Figures 12.7 and 12.8.

Step 2:
Run Contingency Analysis (Figure 12.9).

Step 3:
See Figure 12.10. Check the first section, Contingency Definition (No. 1). Every network element in Figure 12.10's lower section may encounter issues, and these faults are saved in the system memory for future research. To proceed to the next step, click the Execute button (No. 3).

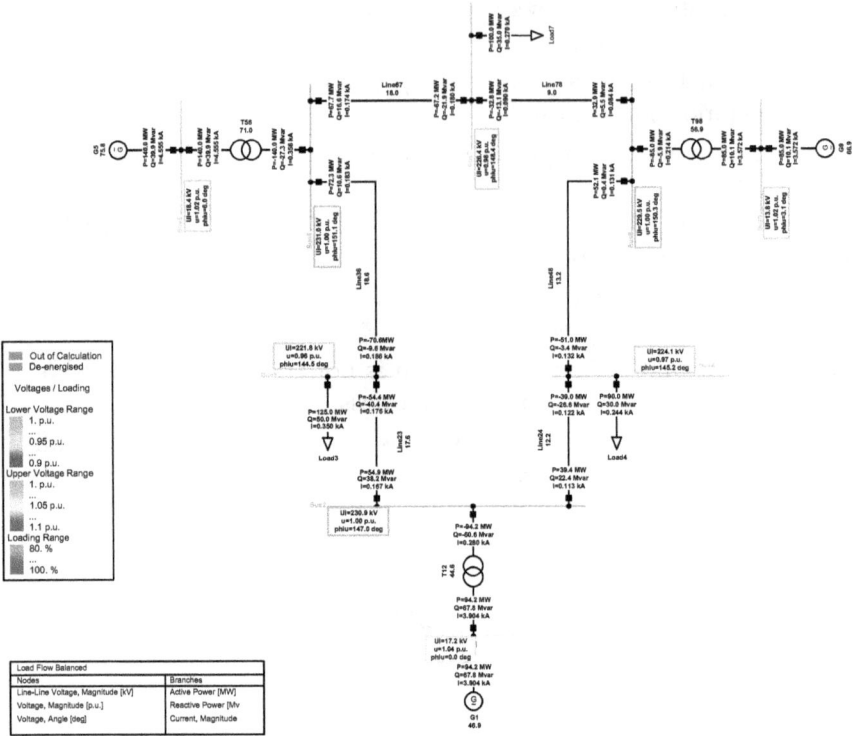

FIGURE 12.7
9-Bus system load flow output (v2021).

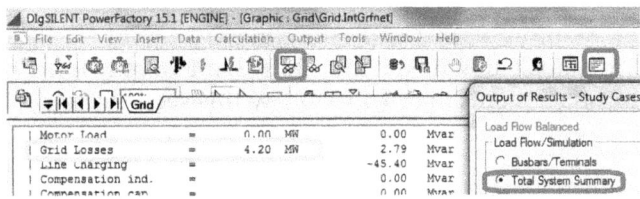

FIGURE 12.8
Grid loss (v15.1).

Step 4:

In this step, Figure 12.11 is seen. Usually, by default, the correct ticks and assumptions are selected. Note the number of contingencies, which is 12. To proceed to the next step, click the Execute button. Following this step, a network divergence error might be

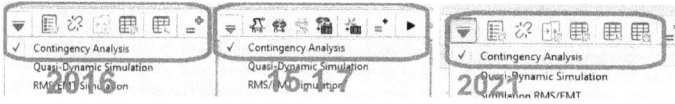

FIGURE 12.9
Contingency analysis toolbox.

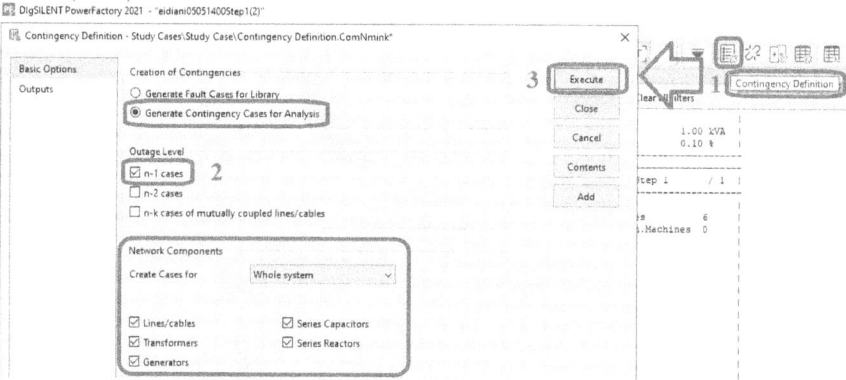

FIGURE 12.10
Contingency definition window.

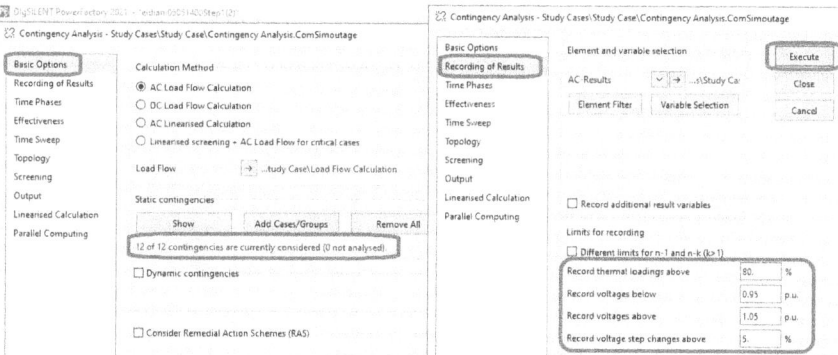

FIGURE 12.11
Contingency analysis window.

seen, indicating that the load flow has diverged in some of the disconnected elements, which is a normal problem.

Step 5:

Now you can get the appropriate output from this section. As shown in Figure 12.12, press the output key to check contingencies.

In the figure shown in Figure 12.3, you can view each of the reports by checking the output. And this section ends.

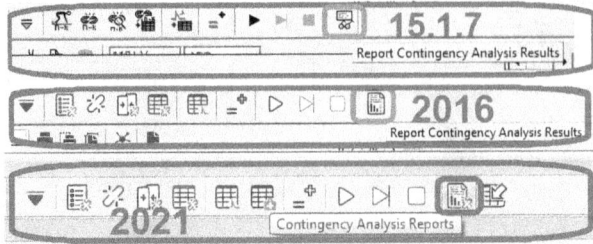

FIGURE 12.12
The output key for checking contingencies.

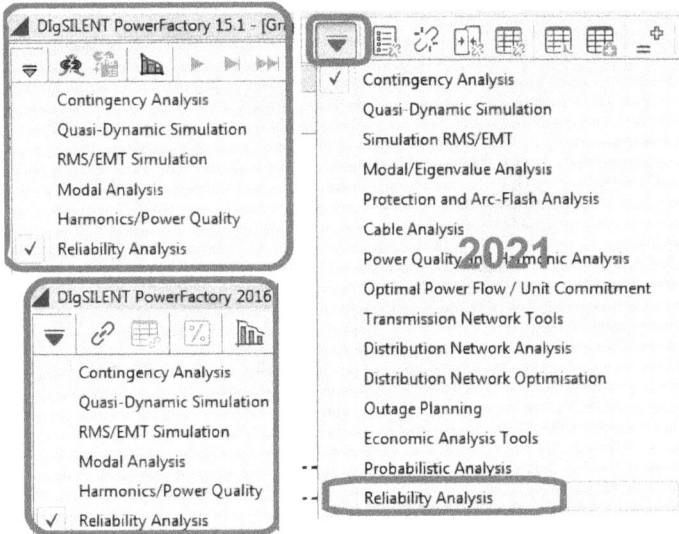

FIGURE 12.13
The Software Reliability Toolbox.

Step 6:

In this section, first enable the Software Reliability Toolbox according to Figure 12.13.

Step 7:

Following the steps in Figure 12.14, you need to enter the busbar type and fill in the reliability parameters. Then, as shown in Figure 12.15, copy the model of one bus and paste it into the other buses. Complete the line and transformer reliability parameters as in Figures 12.14 and 12.15, as in Figures 12.16 and 12.17.

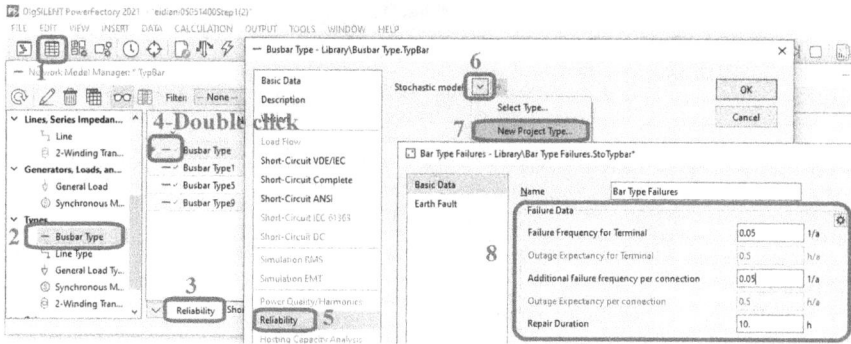

FIGURE 12.14
The busbar type reliability parameters.

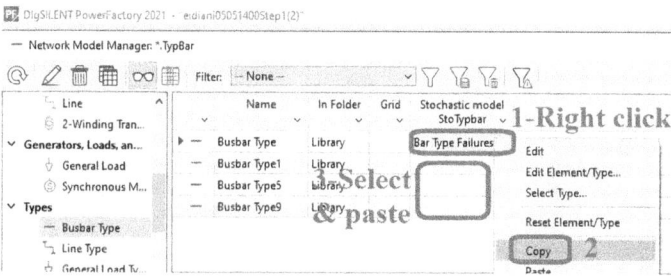

FIGURE 12.15
Copy and paste the reliability model.

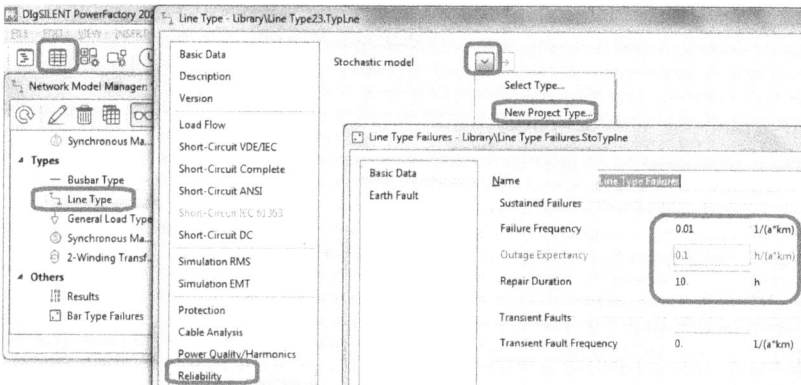

FIGURE 12.16
The line type reliability parameters.

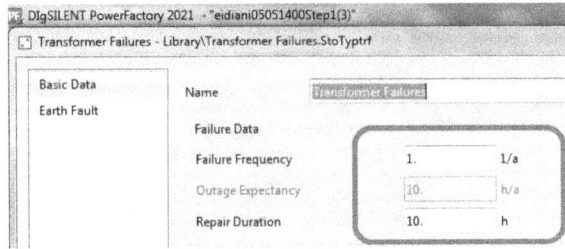

FIGURE 12.17
The transformer type reliability parameters.

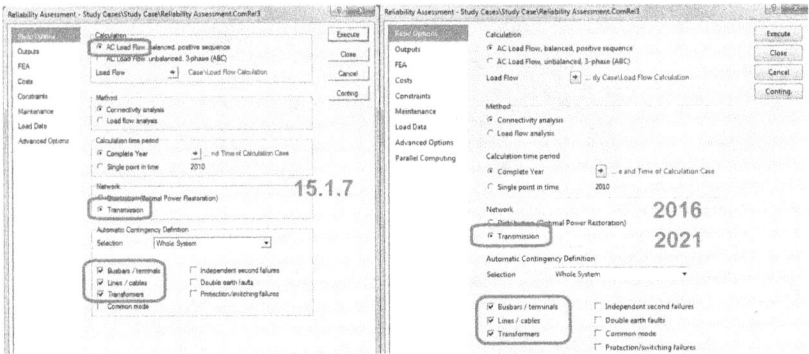

FIGURE 12.18
The reliability assessment window.

Step 8:

Now press the Reliability key (🕭) or ( ) and complete the
opened window as shown in Figure 12.18 and then press the
Execute key. Network Reliability is executed. See the text output.

Further reliability analyses can be carried out using the data
from this section and the preceding section.

12.8 Summary

This chapter focused on using DIgSILENT PowerFactory software to conduct reliability analysis in power systems, emphasizing the importance of
system reliability, which is defined by time, operating conditions, probability, and performance. It introduced the concepts of System Security
and System Adequacy to better understand reliability and discusses key
ideas such as Energy Not Supplied (ENS) and Contingency Analysis. The
appendix guides software usage and potential future work, with references for further reading.

References

1. Eidiani, M., Shanechi, M.H.M., Vaahedi, E., "Fast and accurate method for computing FCTTC (first contingency total transfer capability)," *Proceedings International Conference on Power System Technology*, 2002, vol. 2, pp. 1213–1217. https://doi.org/10.1109/ICPST.2002.1047595
2. Billinton, R. *Power System Reliability Evaluation*, Gordon and Breach, 1970.
3. Grigsby, L.L. *Power Systems*, CRC Press, 2017.
4. Anand, A., Ram, M. *System Reliability Management Solutions and Technologies*, Taylor & Francis Group, 2021.
5. Fuller, J., Obiomon, P., Abood, S.I. *Power System Operation, Utilization, and Control*, CRC Press, 2022.
6. Tjernberg, L.B. *Infrastructure Asset Management with Power System Applications*, CRC Press, 2018.
7. Karami, E., Khalilinia, A., Bali, A., Rouzbehi, K., "Monte-Carlo-based simulation and investigation of 230 kV transmission lines outage due to lightning," *High Voltage*, vol. 5, no. 1, pp. 83–91, 2020.

13

Wind Farm Construction

13.1 Introduction

This chapter uses DIgSILENT PowerFactory software for Wind Farm Construction. Wind farms are increasingly significant as a clean and renewable energy source. By harnessing the wind's kinetic energy, they generate electricity, offering an environmentally friendly alternative to fossil fuel-based power generation.

Ancient Persians are credited with being among the first to harness wind power for practical applications. They developed vertical-axis windmills around 500–900 CE, primarily to grind grain or pump water. These early windmills were an innovative use of wind energy and a significant technological achievement of the time. According to a general estimate, the amount of wind energy that can be transformed into electrical energy on the planet Earth is 72 TW, though this figure can be raised. However, the 20th century saw the development of modern wind turbines, propelled by technological breakthroughs and the expanding demand for clean energy. An important turning point in the evolution of wind power was reached in 1981 when the first commercial wind farm was built in California, USA.

With continuous technological advancements and a growing demand for renewable energy worldwide, wind power seems to have a bright future. Wind power is set to become more prevalent in the world's energy mix as wind turbine efficiency increases and costs keep falling. Particularly in coastal areas, offshore wind farms, which are situated in deeper waters, have enormous potential to produce enormous amounts of clean energy.

Wind power does have its challenges, with one of the primary concerns being its intermittent nature. Variations in wind speed can lead to fluctuations in power generation, making it less predictable and potentially impacting its reliability as a consistent energy source. Batteries and other energy storage devices are being used to address this issue by storing extra energy for use when wind speed is low. The possible effects on wildlife, especially birds and bats, present another difficulty. To reduce these effects, careful site selection and mitigation strategies are necessary. Additionally, some communities may be concerned about the visual impact of wind turbines, which calls for careful planning and public involvement.

DOI: 10.1201/9781003590514-13

The construction and analysis of farms in the DIgSILENT PowerFactory software are covered in this chapter. The analysis is explained first, followed by an explanation of the software tips. How to use the software and the possibility for further work are explained in the appendix. For more detailed and further works, you can refer to the references [1–14] at the end of this chapter.

13.2 DFIG Wind Power Plant

This chapter models the wind farm using only DFIG wind turbines. This is because, as will be discussed below, this kind of power plant is extremely important.

DFIG wind power is an excellent technology due to its numerous advantages. Even when the wind isn't blowing as hard, it allows wind turbines to spin at different speeds, increasing their energy capture capacity. By regulating voltage and power flow, DFIGs also contribute to the stability of the power grid. Because they require smaller power electronic converters than other generator types, they are more cost-effective and use less energy. DFIGs are dependable and flexible because they can be readily connected to the current electrical grid and even operate independently. Additionally, they can remain connected to the grid during a power outage, ensuring a continuous power supply DFIG wind power is a useful instrument for developing a sustainable energy future because of all these benefits.

13.3 Wind Farm Analysis

Consider the 9-bus network from earlier chapters. As seen in Figure 13.1, two 45 MW wind farms have been added to buses 3 and 5 of this network. There are thirty 1.5 MW turbines on each of these two wind farms. The appendix explains how this farm was constructed.

Line 3–6 is now experiencing a symmetrical three-phase short circuit, in which the short circuit occurs in second (1) and is cleared in (1.2) seconds. Keep in mind that there are two 45 MW wind farms in this transient stability analysis, for a total of sixty 1.5 MW DFIG wind turbines in a nine-bus network. The generator angles' output is shown in Figure 13.2. The maximum generator angle is 119°. As you can see, the existence of 60 wind farms contributes to the lengthy computation time.

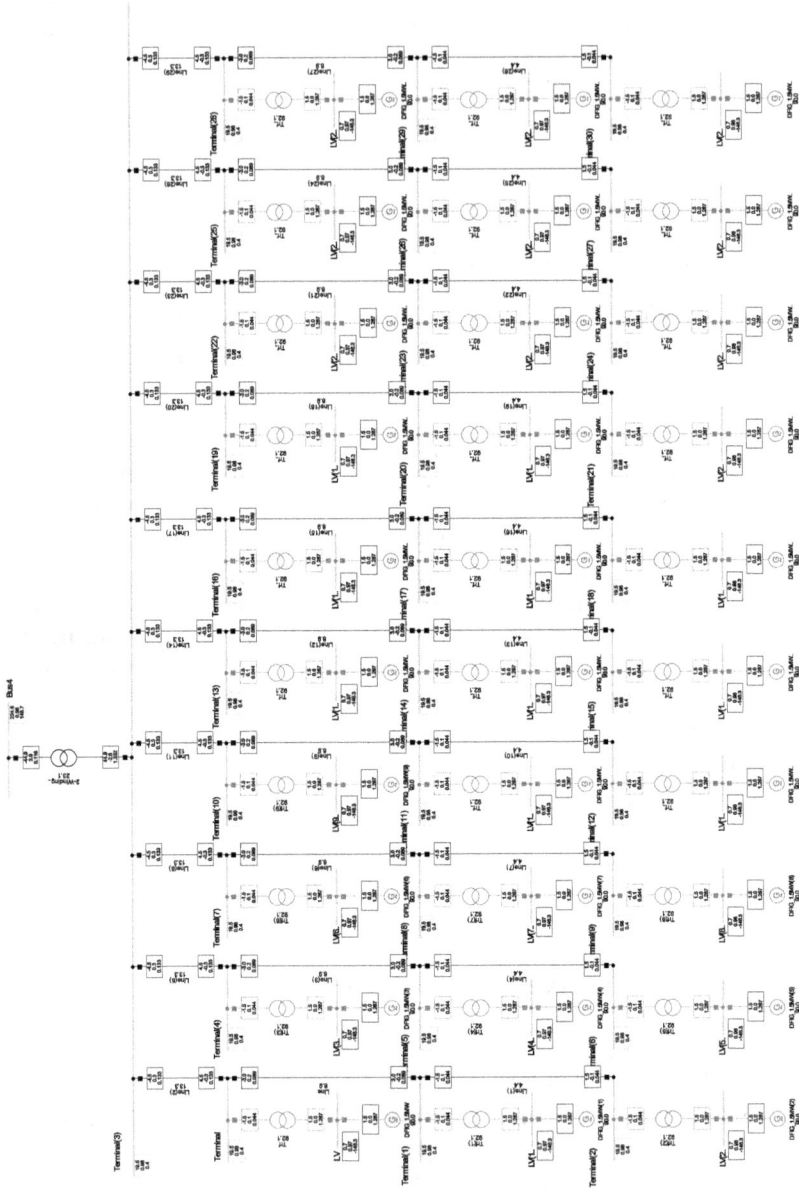

FIGURE 13.1
A 45 MW farm consisting of 30 1.5 MW turbines.

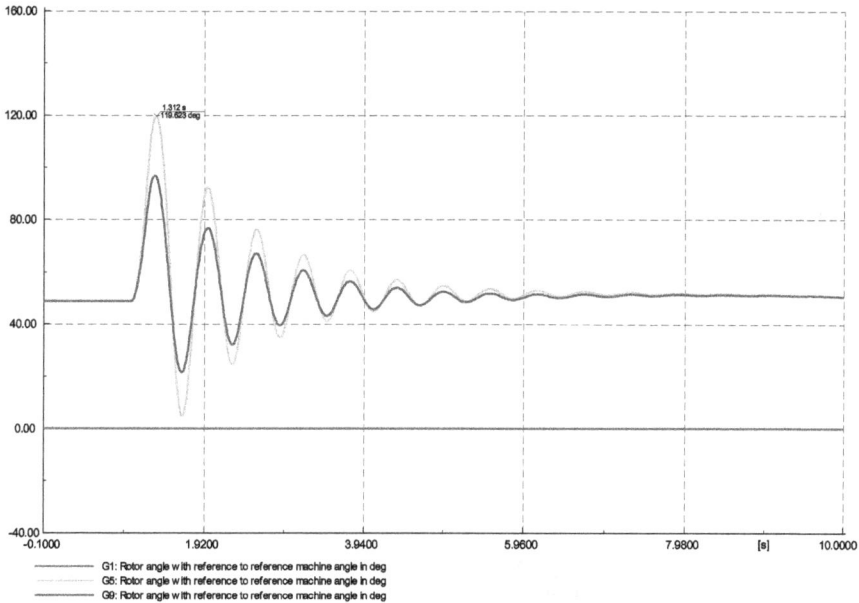

FIGURE 13.2
Transient stability output with 60 wind turbines with high computation time.

In this phase, the transient simulation is conducted again with the entire wind farm taken out of service. There is a significant reduction in computation time. The transient stability output is shown in Figure 13.3, with a maximum angle of 95°.

This section makes use of parallel generators. At one point, a farm uses 30 parallel wind turbines rather than 30 wind turbines. Figure 13.4 illustrates these two instances. Figure 13.5 displays the model's transient stability output.

Although the maximum angle remains at 119° and the figures are similar, the simulation calculation speed is extremely high. Consequently, it can be said that using parallel wind generators in wind farms speeds up computations and has little effect on transient stability.

The appendix contains information that explains how to use DIgSILENT software to build a wind farm step-by-step. Additionally, readers can perform further analysis on the wind farm by uploading the wind farm file provided with the book.

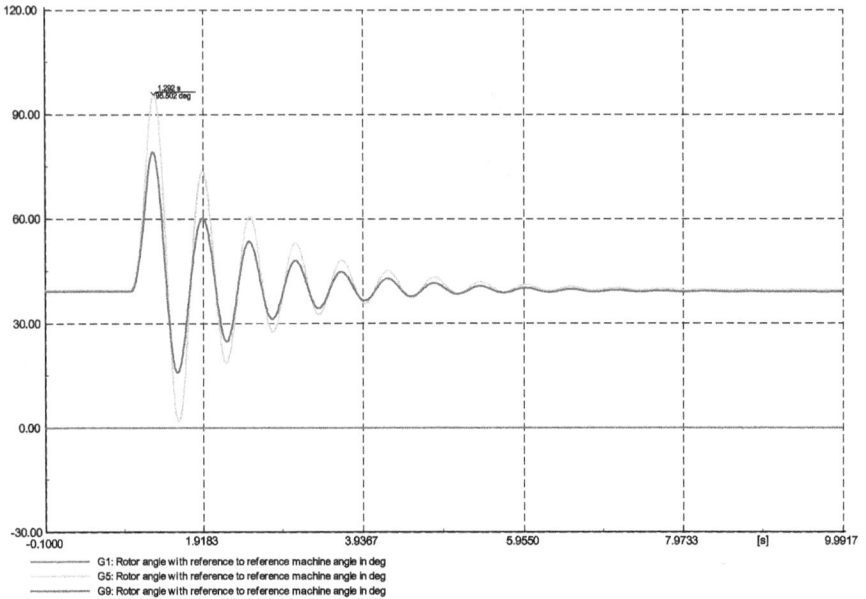

FIGURE 13.3
Transient stability output without 60 wind turbines with low computation time.

13.4 Two-Choice Questions (Yes/No)

1. Wind power plant simulations can help predict energy output.
2. Wind farm simulations are not useful for optimizing turbine placement.
3. Atmospheric pressure does not significantly influence wind power generation.
4. Wind direction can affect the power output of a wind farm.
5. Wind speed is a critical input parameter for wind power simulations.
6. Terrain roughness has no impact on wind speed and direction.
7. Wind turbine wake effects can reduce the overall power output of a wind farm.
8. Wind shear does not affect the power output of a wind turbine.
9. Power curve data is not necessary for wind power simulations.
10. Wind power simulations can help assess the impact of different turbine control strategies.

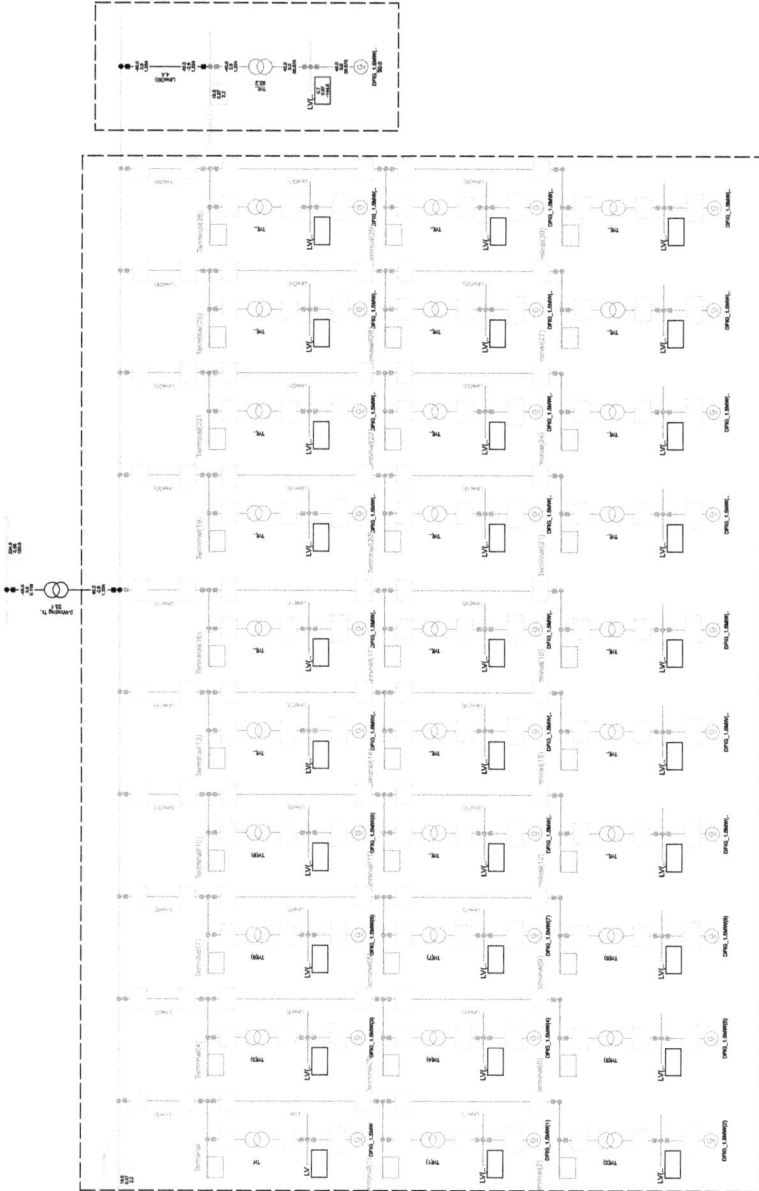

FIGURE 13.4
Compact and parallel model of a 45 MW wind farm.

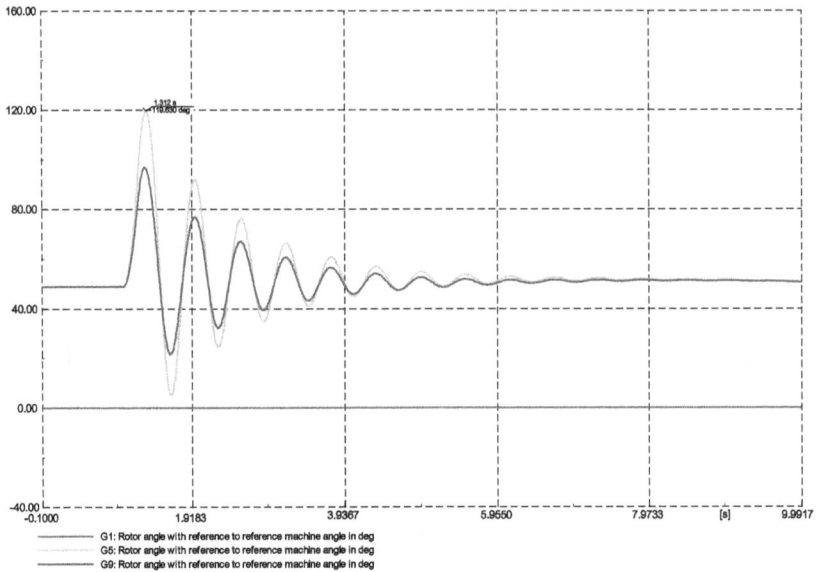

FIGURE 13.5
Transient stability output with compact and parallel model of a 45 MW wind farm with low computation time.

11. Computational Fluid Dynamics (CFD) can be used to simulate wind flow around wind turbines.

12. Wind power simulations can be used to estimate the levelized cost of energy (LCOE).

13. Climate change does not affect long-term wind power predictions.

14. The capacity factor of a wind farm is always 100%.

15. Wind power simulations can help identify potential grid integration issues.

16. Wind power simulations are not useful for evaluating the environmental impact of wind farms.

17. Wind speed data is always accurate and reliable.

18. The power curve of a wind turbine remains constant over its lifetime.

19. Wind power simulations can help assess the impact of different turbine technologies.

20. Wind resource assessment is a crucial step in wind farm development.

21. Wind power simulations can help optimize the maintenance schedule for wind turbines.

22. Wind power simulations can help identify suitable locations for wind farms.

23. Wind power simulations are only useful for large-scale wind farms.

24. The output of a wind power simulation is always deterministic.

25. Wind power simulations are not useful for assessing the impact of extreme weather events.

26. The cost of wind power simulations is negligible.

27. Wind power simulations can help optimize the operation of wind farms.

28. Wind power simulations can be used to evaluate the impact of different meteorological models.

29. Wind power simulations can help assess the impact of different financing models.

30. Wind power simulations are not useful for assessing the impact of grid codes and standards.

31. Wind power simulations can help identify potential noise pollution issues.

32. The impact of bird and bat mortality on wind farms is not considered in simulations.

33. The accuracy of wind power simulations is independent of the quality of input data.

34. Wind power simulations can help assess the impact of different maintenance strategies.

35. Wind power simulations can help assess the impact of different turbine spacing configurations.

36. Wind power simulations can help assess the impact of different energy storage technologies.

37. The impact of offshore wind farms on marine ecosystems is not considered in simulations.

38. Wind power simulations are not useful for evaluating the impact of different grid connection scenarios.

39. Wind power simulations can help identify potential risks and uncertainties associated with wind energy projects.

40. The cost of wind power simulations is decreasing over time.

13.4.1 Key Answers to Two-Choice Questions

Yes (True)	1, 4, 5, 7, 10–12, 15, 19–22, 27–29, 31, 34–36, 39, 40
No (False)	2, 3, 6, 8, 9, 13, 14, 16–18, 23–26, 30, 32, 33, 37, 38

13.5 Appendix, Wind Farm Construction in DIgSILENT

This section requires a basic understanding of DIgSILENT Power Factory software. You must download the file (Chapter8.pfd) from the book's end-of-book attachments to follow this section. Of course, if you need the ready file for analysis, use the files (Chapter13-1.pfd) and (Chapter13-2.pfd). We'll give you a quick overview of how to construct wind farms in this section. It is best to have some knowledge about the two types of wind turbines (dual-feed and squirrel cage) before you begin.

The network under study is the same as the previous chapters, with losses of 9 busbars (4.2 MW).

Step 1:
Create a new graphics **Grid** page to build the wind farm, as shown in Figure 13.6.

Step 2:
Go back to the main page **Grid** right-click on **Bus 4** and select **Copy**.

Go to the wind page. Right-click and select **Paste Graphically Only** to make a copy of bus 4 on the new wind page. In this case, only a graphical copy of bus 4 has been created on the wind farm page.

Step 3:
Place a 20 kV terminal on the wind page. (No. 1 Figure 13.7). By clicking the **General Template** button (No. 2 Figure 13.7), select the DFIG-WTG-1.5 MW wind power plant and connect it to the 20 kV terminal. (Steps 2–4). This Figure is slightly different from other versions. See Figure 13.8.

Step 4:
Complete the Stochastic Model of the wind generator by double-clicking on the generator (DFIG) in the opened window, as shown in Figure 13.9.

Step 5:
On the same **Generation Adequacy** page as in Figure 13.10, add the wind mean and variance or Beta.

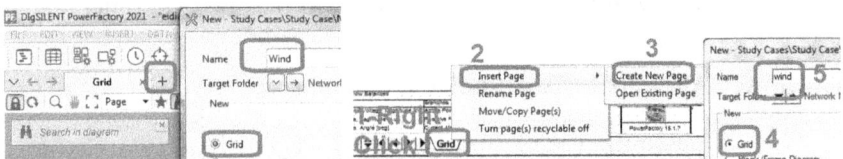

FIGURE 13.6
Create a new graphics Grid page.

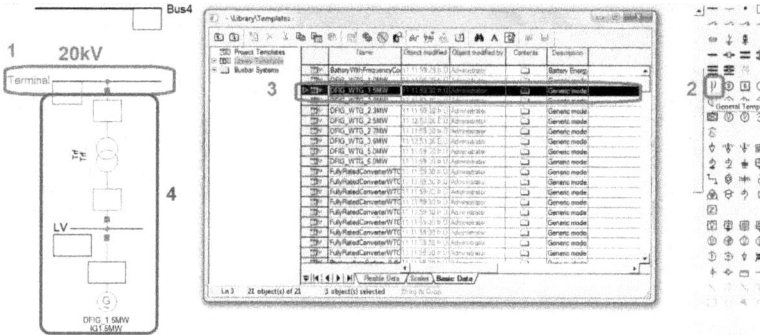

FIGURE 13.7
Create one DFIG wind turbine (v15.1).

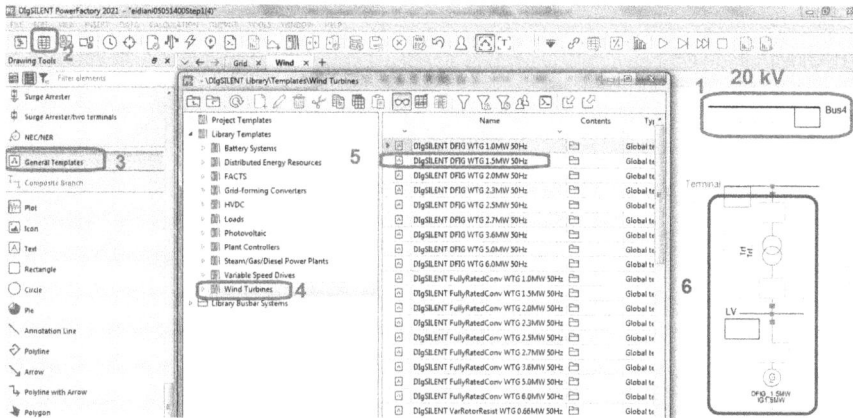

FIGURE 13.8
Create one DFIG wind turbine (v2021).

Step 6:

On the same **Generation Adequacy** page as in Figure 13.11, complete the Wind Power Curve.

Step 7:

In this section, select the built generator connected to the 20 kV bus, then right-click on it, select **Define Template**, and name it **New-DFIG** as shown in Figure 13.12.

Step 8:

This simple network is visible in the **General Template**. Double-click on it and click Pack as shown in Figure 13.13. This built system is ready.

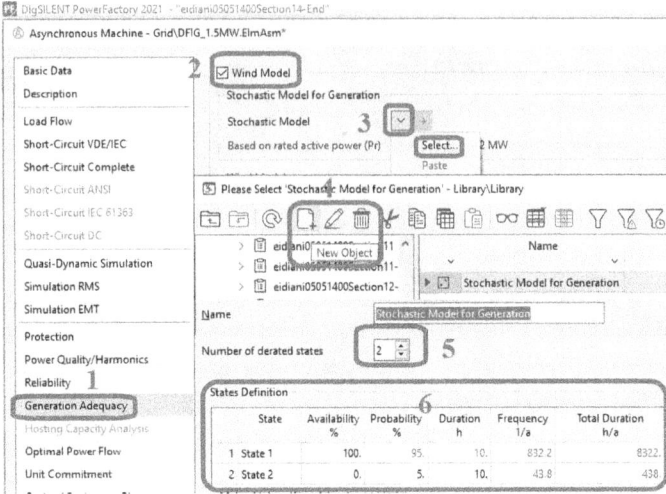

FIGURE 13.9
Stochastic Model window.

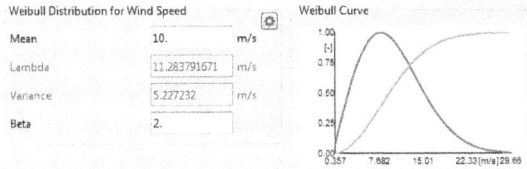

FIGURE 13.10
Weibull distribution for wind speed window.

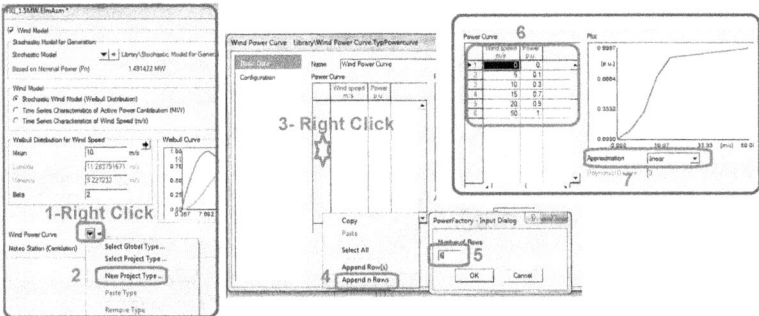

FIGURE 13.11
Wind power curve window.

Step 9:

9-1 In the General Template, select New-DFIG and double-click on the page.

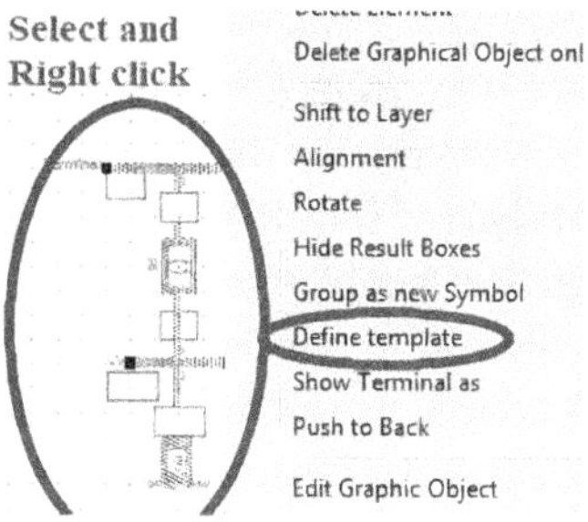

FIGURE 13.12
Define template window.

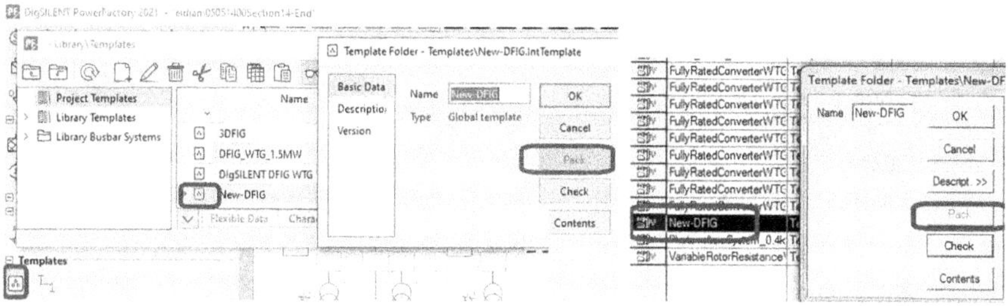

FIGURE 13.13
Packing the general template network.

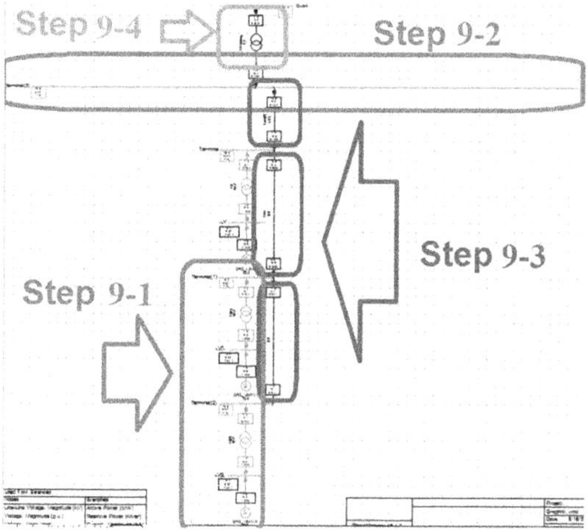

FIGURE 13.14
Building a wind farm with three generators in series.

9-2 Create a large 20 kV bus at the top of the page (See Figure 13.14).

9-3 Connect three one-kilometer 20 kV lines between the three constructed parts. Assume their resistance and reactance are 0.01 Ω.

If there is not enough space on the page, right-click on a white area of the page, select Drawing Format, and select A3 paper.

9-4 Now place a transformer between the 20 kV bus and bus 4 with a voltage of 230 kV (Figure 13.14). Make its type by copying the generator transformer Winding Transformer Type1. Just convert the voltage from 18 to 20 kV. (Figure 13.15). Run the load flow and you should not see any error.

Step 10:

Now select the top transmission line (without 20 kV bus) to the lowest generator and right click on it and select Define template and name it 3DFIG.

Step 11:

The three-generator network is built in the General Template. Double-click on it and, as before, click Pack. This built system is ready.

Step 12:

In the General Template, select 3 DFIG and check 10 times (create 10 sample networks) so that the top line is automatically connected to the large 20 kV bus. Run load flow. You should have created a figure similar to Figure 13.1. The wind farm must generate 44.95 MW of power without failure.

Step 13:

Now, right-click on this 45 MW wind farm and select the entire farm (without bus 4). After choosing Define Template, give it the name Wind-Farm-45 MW. This farm can also be utilized in other buses or networks. Remember to bring the pack key.

On bus 3, construct a second 45 MW wind farm and run the load flow. There should be 3.8 MW of network losses. Building a farm with thirty generators now takes ten seconds!

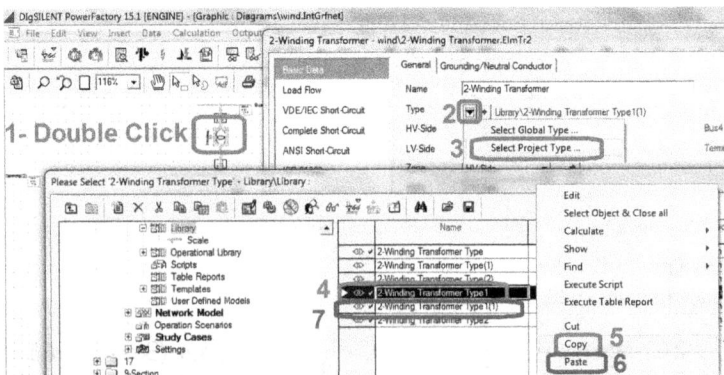

FIGURE 13.15
Copying the wind farm transformer type from the type of another transformer.

Step 14: (Build parallel wind generators)

This section teaches how to build parallel wind generators. As shown in Figure 13.16, with the help of General Template and the NEW-DFIG network, build a model of 30 parallel lines (No. 3), 30 parallel transformers (No. 4) and 30 parallel generators (number 5). The main transformer and the main bus are back in service (No.s 1 and 2).

Double-click on the wind generator (No. 7) and look for (PQ_tot) in the Plant Model section. Double-click and set the rated power to (45) MW (30 × 1.5) (No.s 7–10 in Figure 13.16).

Run load flow. The output power should be 44.96 MW as before. Do the same for the wind farm in Bus 3.

Step 15: (Plotting the two outputs in one figure)

Plotting the two outputs before and after the change in one figure is what you will learn to do in this section. Just copy and paste "All calculations" following the initial simulation to accomplish this, as illustrated in Figure 13.17. "All calculations(1)" is where the first simulation is stored. You can now run the simulation and switch the network. "All calculations" is where the data is stored.

Step 16:

Double-click now to create two instances of the generator angles in the simulation output page, as illustrated in Figure 13.18. The first simulation is represented by one, and the second by the other.

Figure 13.19 shows the transient stability of the system with and without wind farms.

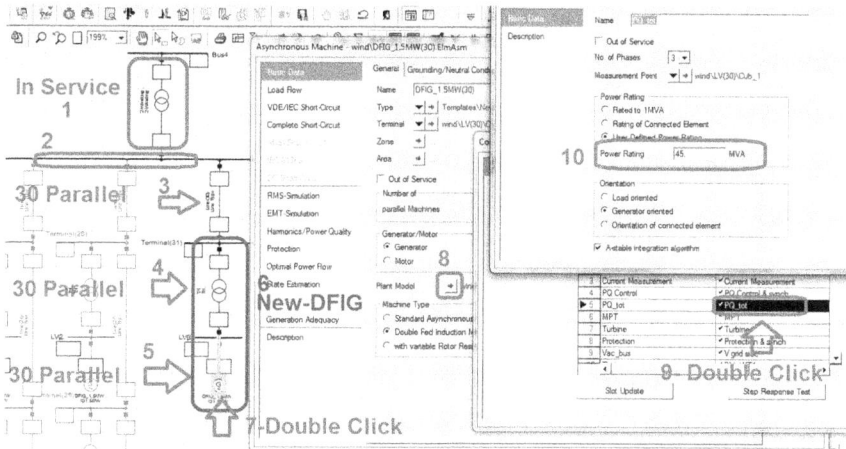

FIGURE 13.16
Construction of a 30-generator compact wind farm.

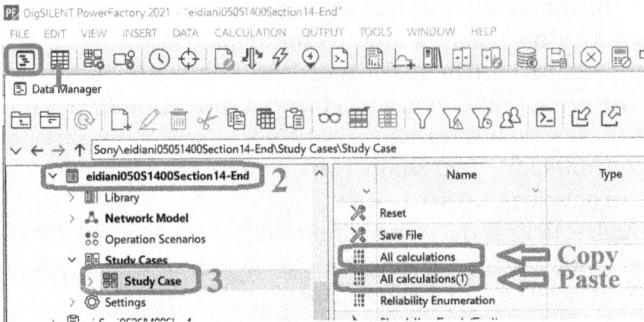

FIGURE 13.17
Copy and paste "All calculations".

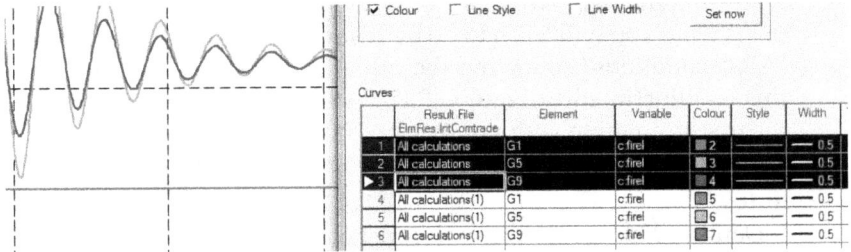

FIGURE 13.18
Adding two information before and after changes to the network.

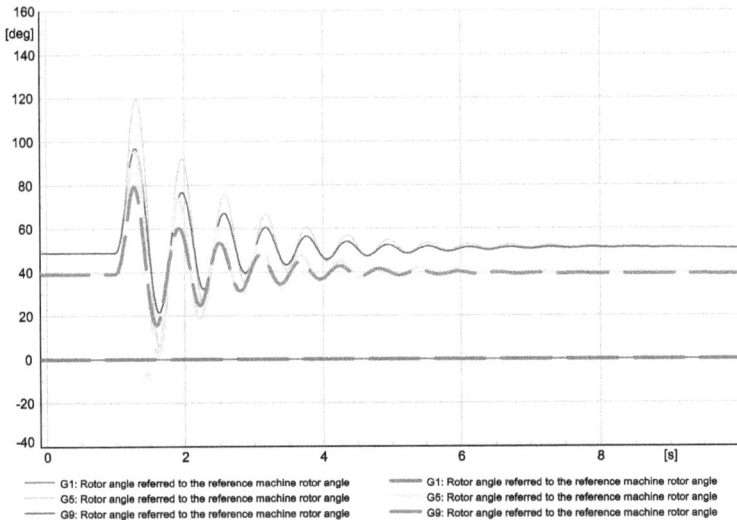

FIGURE 13.19
Transient stability output with and without 60 parallel compact wind turbines in one graph (v2021).

13.6 Summary

This chapter discussed the use of DIgSILENT PowerFactory software for the construction and analysis of wind farms, highlighting the growing significance of wind energy as a clean and renewable power source. It traces the historical development of wind power, noting the first commercial wind farm in California in 1981 and the potential for offshore wind farms to generate substantial energy. While acknowledging the challenges of wind energy, such as its intermittent nature and environmental impacts, the chapter emphasizes ongoing technological advancements and the importance of careful planning and community engagement. It concludes with guidance on using the software and references for further exploration.

References

1. Eidiani, M., Mahnani, S. "Optimally and independent planned microgrid with solar-wind and biogas hybrid renewable systems by HOMER," *Majlesi Journal of Electrical Engineering*, 2023, 17(2), pp. 153–158, https//doi.org/10.30486/mjee.2023.1974843.1021
2. Eidiani, M., Asghari Shahdehi, N., Zeynal, H. "Improving dynamic response of wind turbine driven DFIG with novel approach," *2011 IEEE Student Conference on Research and Development*, 2011, pp. 386–390, https//doi.org/10.1109/SCOReD.2011.6148770
3. Ghardashi, G., Gandomkar, M., Majidi, S., Eidiani, M., Dadfar, S. "Accuracy and speed improvement of microgrid Islanding detection based on PV using frequency-reactive power feedback method," *2022 International Conference on Protection and Automation of Power Systems (IPAPS)*, 2022, pp. 1–8, https//doi.org/10.1109/IPAPS55380.2022.9763190
4. Eidiani, M., Ghavami, A. "New network design for simultaneous use of electric vehicles, photovoltaic generators, wind farms and energy storage," *2022 9th Iranian Conference on Renewable Energy & Distributed Generation (ICREDG)*, 2022, pp. 1–5, https//doi.org/10.1109/ICREDG54199.2022.9804534
5. Eidiani, M., Zeynal, H., Ghavami, A., Zakaria, Z. "Comparative analysis of mono-facial and bifacial photovoltaic modules for practical grid-connected solar power plant using PVsyst," *2022 IEEE International Conference on Power and Energy (PECon)*, 2022, pp. 499–504, https//doi.org/10.1109/PECon54459.2022.9988872
6. Eidiani, M., Zeynal, H., Zakaria, Z., Shaaban, M. "Analysis of optimization methods applied for renewable energy integration," *2023 IEEE 3rd International Conference in Power Engineering Applications (ICPEA)*, 2023, p. 1570861924.

7. Eidiani, M. "Applying optimization techniques to develop a renewable energy supply map," In: Fathi, M., Zio, E., Pardalos, P.M. (eds) *Handbook of Smart Energy Systems,* Springer, Cham, 2022. https://doi.org/10.1007/978-3-030-72322-4_61-1

8. Zand, Z., Ghahri, M.R., Majidi, S., Eidiani, M., Nasab, M.A., Zand, M. "Smart grid and resilience," In: Fathi, M., Zio, E., Pardalos, P.M. (eds) *Handbook of Smart Energy Systems,* Springer, Cham, 2022. https://doi.org/10.1007/978-3-030-72322-4_178-1

9. Momen, A., Hekmati, A., Majidi, S., Zand, Z., Zand, M., Nikoukar, J., Eidiani, M. "Energy harvesting for smart energy systems," In: Fathi, M., Zio, E., Pardalos, P.M. (eds) *Handbook of Smart Energy Systems,* Springer, Cham, 2022. https://doi.org/10.1007/978-3-030-72322-4_12-1

10. Eidiani, M. "Integration of renewable energy sources," In: Fathi, M., Zio, E., Pardalos, P.M. (eds) *Handbook of Smart Energy Systems,* Springer, Cham, 2022. https://doi.org/10.1007/978-3-030-72322-4_41-1

11. Eidiani, M., Zeynal, H., Ghavami, A., Zakaria, Z. "Performance of bifacial solar system for 1MW grid-connected solar power plant in Mashhad, Iran: design, simulation and economic analysis," *2024 9th International Conference on Technology and Energy Management, ICTEM 2024,* 2024.

12. Eidiani, M., Ghavami, A., Zeynal, H., Zakaria, Z. "Performance of mono-facial and bifacial solar system for 5MW grid-connected solar power plant with fixed tilt and tracking: design, simulation and investigating the effect of temperature," *2024 28th International Electrical Power Distribution Conference, EPDC 2024,* 2024.

13. Raza, A., Younis, M., Liu, Y., Altalbe, A., Rouzbehi, K., Abbas, G., "A multi-terminal HVdc grid topology proposal for offshore wind farms," *Applied Sciences,* vol. 10, no. 5, pp. 1833, 2020.

14. Yazdi, S. H., Milimonfared, J., Fathi, S. H., Rouzbehi, K., "Analytical modeling and inertia estimation of VSG-controlled Type 4 WTGs: Power system frequency response investigation," *International Journal of Electrical Power & Energy Systems,* vol. 107, pp. 446–461, 2019.

14

Large Network Simulations

14.1 Introduction

This chapter explores large network simulations using DIgSILENT PowerFactory software, providing both the necessary foundational information and practical applications. Several large-scale networks are analyzed, beginning with the Texas grid, which features over 2,000 buses, 2,300 lines, and 500 generators. Subsequent analyses cover networks such as the Frankfurt (Germany) 400 kV Grid, the Tübingen (Germany) 110 kV Grid, a 400 MW offshore wind farm connected via an HVDC line, the Khorasan province (Iran) Grid, and the IEEE 118-BUS (modified) system.

Each network is examined under various operational scenarios, including transient stability, contingency ranking, and frequency analysis. Simulation files for each network are provided in the book's appendix, enabling readers to replicate the results in PowerFactory. Detailed analyses of these networks are presented in the sections that follow, with references [1–7] available for further exploration and advanced studies.

14.2 Texas_USA Grid Example

This network has 10 voltage levels from 13.2 to 500 kV and is also composed of 8 sub-networks, as shown in Figure 14.1 and Table 14.1.

You can simply replicate the simulations in the attached file by loading and importing it, as illustrated in Figure 14.2. Two of its well-known sections are rebuilt in this section. Texas grid is subjected to a 10-second RMS simulation with and without PSS in the first section. It was expected that PSS would significantly lessen the grid fluctuations.

We then look at the network's contingency analysis. Out of all the line, load, transformer, and generator contingencies, there are 3750 static contingencies. The worst-case scenario for the ODESSA_6–115 kV bus voltage drops of 0.102 per unit is the 510 contingency, or loss of line lne_1024_1090_1, as seen in Figure 14.3. The contingencies are therefore poorly ranked. Every regional power company performs a daily and

DOI: 10.1201/9781003590514-14

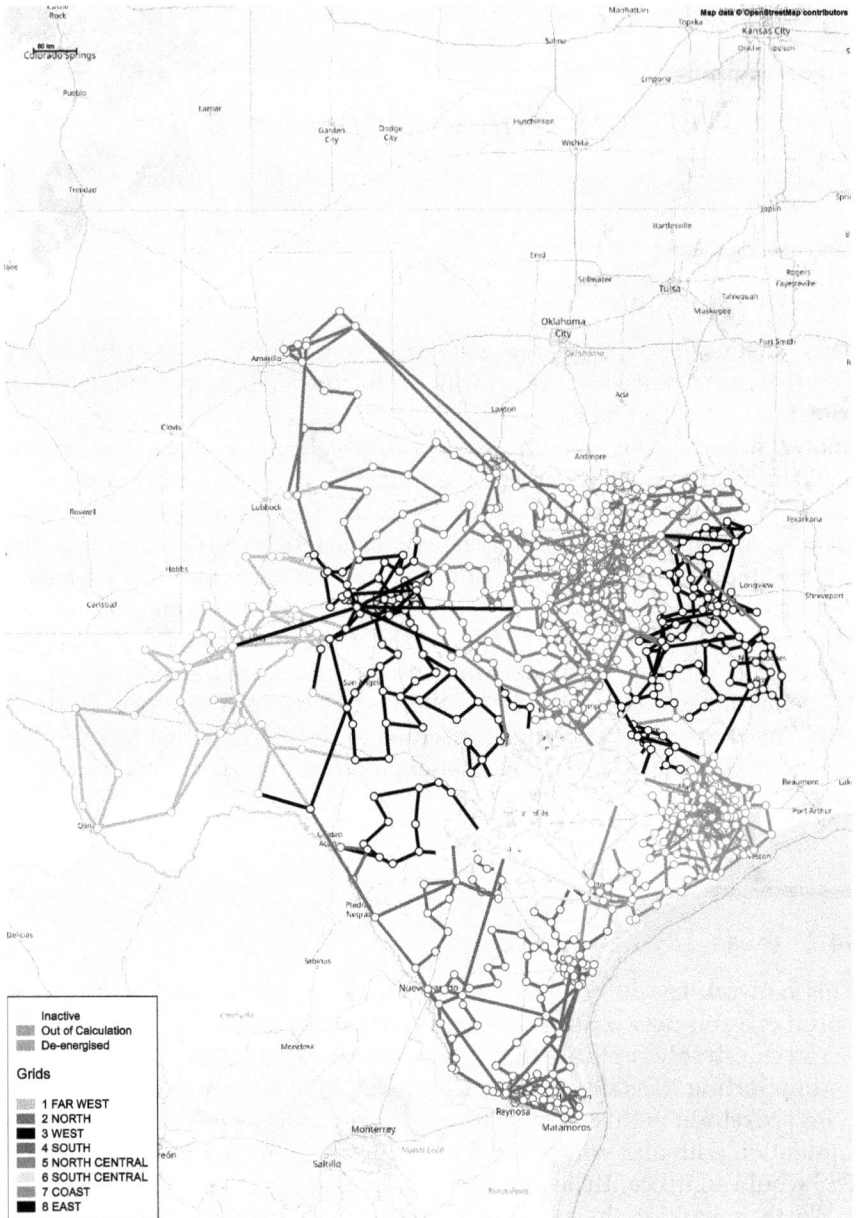

FIGURE 14.1
The Texas grid example 2000-bus model.

TABLE 14.1

Total System Summary of Texas Grid

Number of Substations 1693	Number of Busbars 2000	Number of Loads 1350	Number of SVS 152
Number of Lines 2345	Number of 2-w Trfs. 861	Number of syn. Machines 432	

Generation=68692.49 MW 10793.80 Mvar 69535.34 MVA

External Infeed=0.00 MW 0.00 Mvar 0.00 MVA

Load P(U)=67109.24 MW 19014.28 Mvar 69750.93 MVA

Grid Losses=1583.25 MW 9307.71 Mvar	Line Charging=−11986.63 Mvar
Compensation ind.=462.21 Mvar	Compensation cap.= −17990.41 Mvar
Installed Capacity=104813.14 MW	Spinning Reserve=12926.96 MW

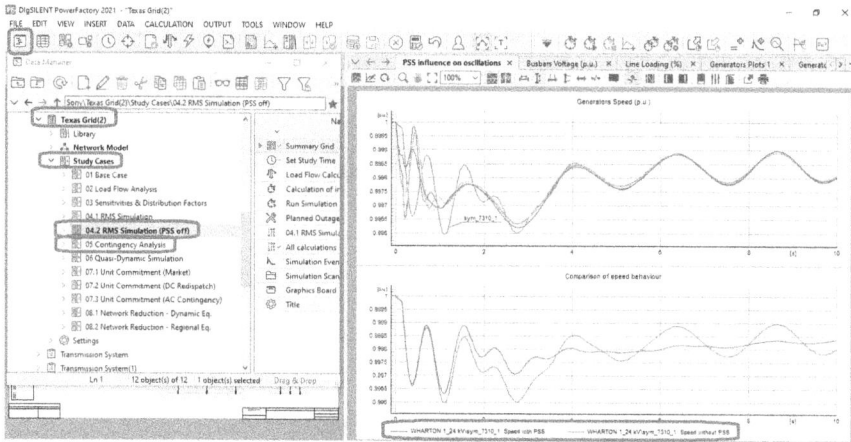

FIGURE 14.2
Simulation with and without PSS.

routine task called "Contingency Ranking," in which $(n-1)$ contingencies are examined for every hourly and daily change in the network. There is no need for extra preparations for the remaining safe contingencies if the network is protected from the worst.

14.3 Frankfurt_Germany 400 kV Grid Example

Figure 14.4 depicts a portion of Frankfurt, Germany's 400 kV transmission network. The pertinent file can be imported from the book's appendix in order to analyze this network. About 400 buses and terminals, 25 lines

Contingency Analysis Report: Voltage steps

Study Case: 05 Contingency Analysis
Result File: Contingency Analysis AC

Max. voltage step 0.030
Min.Voltage Limit: 0.95
Max. Voltage Limit: 1.05

	Component	Branch, Substation or Site	Voltage Step [p.u.]	Voltage Base [p.u.]	Voltage Min./Max. [p.u.]	Contingency Number	Contingency Name	Voltage Step [0.000 p.u. - 0.102 p.u.]
1	ODESSA 6 0	ODESSA 6_115 kV	0.102	0.994	0.893	510	_lne_1024_1090_1	
2	JACKSONVILLE 1 2	JACKSONVILLE 1_115 kV	0.096	1.021	0.925	489	_trf_8113_8114_1 [JACKSONVILLE 1]	
3	KENEDY 1	KENEDY_115 kV	0.093	1.007	0.914	329	_trf_6263_6264_1 [KENEDY]	
4	FREEPORT 1 2	FREEPORT 1_115 kV	0.083	1.014	0.932	381	_trf_7321_7322_1 [FREEPORT 1]	
5	BRENHAM 2	BRENHAM_115 kV	0.081	1.006	0.925	325	_trf_6063_6064_1 [BRENHAM]	
6	NOTREES 0	NOTREES_115 kV	0.077	0.991	0.915	510	_lne_1024_1090_1	
7	JACKSONVILLE 2 0	JACKSONVILLE 2_115 kV	0.076	1.010	0.934	489	_trf_8113_8114_1 [JACKSONVILLE 1]	
8	FREEPORT 3 0	FREEPORT 3_115 kV	0.075	1.006	0.931	381	_trf_7321_7322_1 [FREEPORT 1]	
9	PETTUS 0	PETTUS_115 kV	0.061	0.996	0.935	329	_trf_6263_6264_1 [KENEDY]	
10	MOUNT PLEASANT 2 1	MOUNT PLEASANT 2_115 kV	0.061	1.002	0.942	493	_trf_8105_8106_1 [MOUNT PLEASANT 2]	
11	PALESTINE 1 0	PALESTINE 1_115 kV	0.056	1.020	0.963	489	_trf_8113_8114_1 [JACKSONVILLE 1]	
12	RUSK 0	RUSK_115 kV	0.051	1.014	0.963	489	_trf_8113_8114_1 [JACKSONVILLE 1]	
13	CHAPPELL HILL 0	CHAPPELL HILL_115 kV	0.051	1.004	0.953	325	_trf_6063_6064_1 [BRENHAM]	
14	ABILENE 7 0	ABILENE 7_115 kV	0.045	0.995	0.950	537	_lne_3032_3094_1	
15	DENTON 1 1	DENTON 1_161 kV	0.043	0.997	0.955	314	_trf_5164_5165_1 [DENTON 1]	
16	VICTORIA 2 2	VICTORIA 2_115 kV	0.042	1.006	0.964	424	_trf_7043_7044_1 [VICTORIA 2]	

FIGURE 14.3
Contingency analysis report for Texas grid.

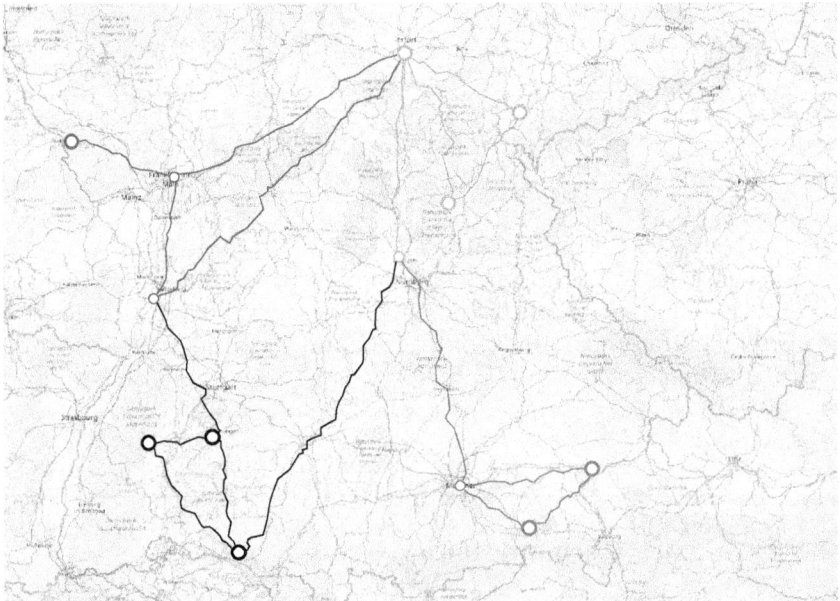

FIGURE 14.4
A 400 kV transmission system.

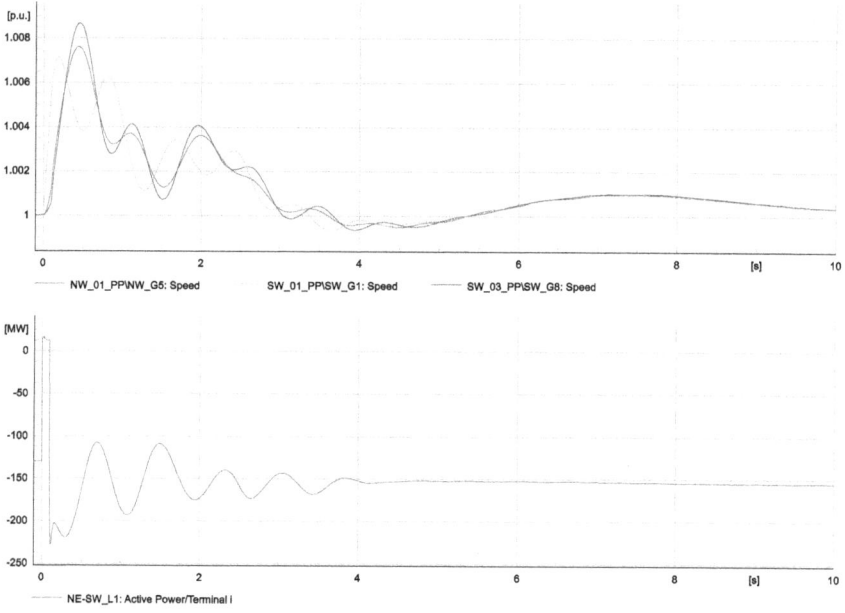

FIGURE 14.5
Network parameter output for 0.1 second fault.

and transformers, and 18 generators make up the model of this network. This section reviews the PV curve, quasi-dynamic simulation analysis, and transient stability analysis.

Initially, we conduct transient stability analysis for a line interruption at 0.1 seconds and a short circuit at 50% of the line (SW_L3). Figure 14.5's output shows a normal trend. This figure illustrates a normal oscillation between the transmitted power and the generators' angular speed. The output of the network parameters for a fault of 1 second is shown in Figure 14.6. Figure 14.7 shows the presence time of the **UnderVoltage** and **OverFrequency** controllers, which prevent network instability if the short circuit time is set to 1 second.

The undervoltage relays on the solar modules run at 0.21, 0.675, and 1.235 seconds, as shown in Figure 14.7. Additionally, the overfrequency relay on the solar module SE_PV1 runs at 1.21 seconds. Despite the lengthy short circuit time, the controller managed to maintain grid stability.

All load and generation data for a day, a month, or a year can be entered discretely in a Quasi-Dynamic Simulation analysis. Figure 14.8 illustrates the process of entering the data. Examining the load flow data, voltage magnitude, power, angle, etc. in one or more curves for a given period is possible by entering the data. The voltage and power for a day are displayed in Figures 14.9 and 14.10. This kind of simulation is crucial for network operation and peak point identification.

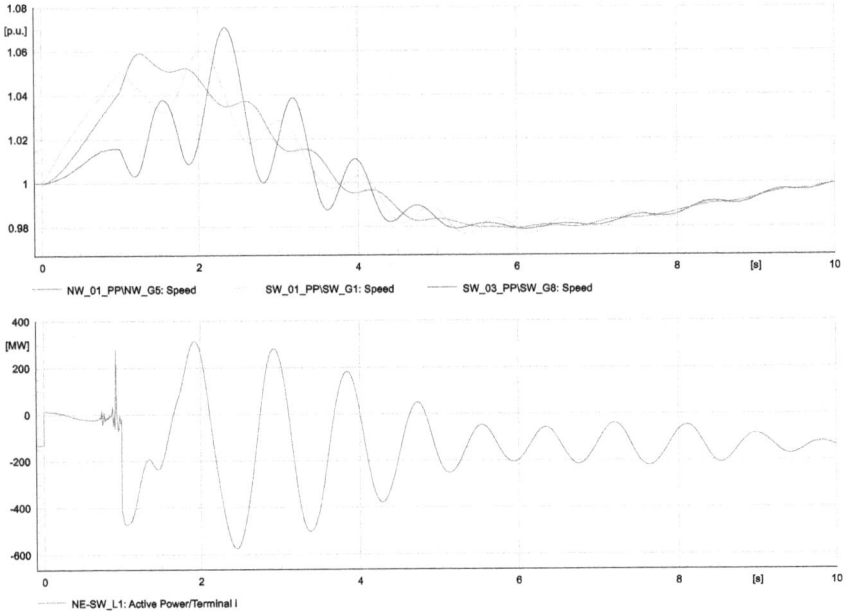

FIGURE 14.6
Network parameter output for 1 second fault.

Name	Time	Object
Short-Circuit Event	0.	SW_L3
UnderVoltage	0.21	SW_PV1
UnderVoltage(1)	0.21	SW_PV2
UnderVoltage(2)	0.675	NW_PV1
Switch Event	1.	SW_L3
OverFrequency	1.21	SE_PV1
UnderVoltage(3)	1.235	SE_PV1

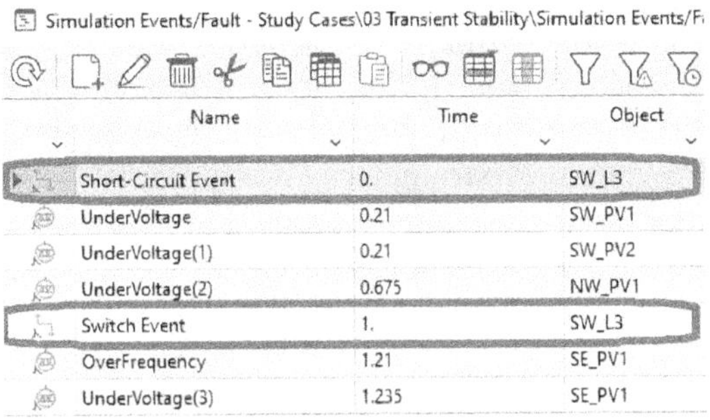

FIGURE 14.7
The UnderVoltage and OverFrequency controllers.

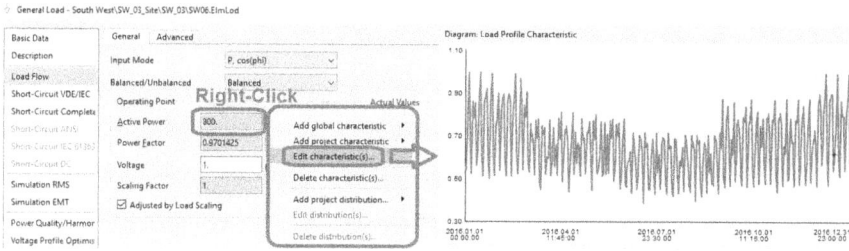

FIGURE 14.8
One year of data entry for load and production.

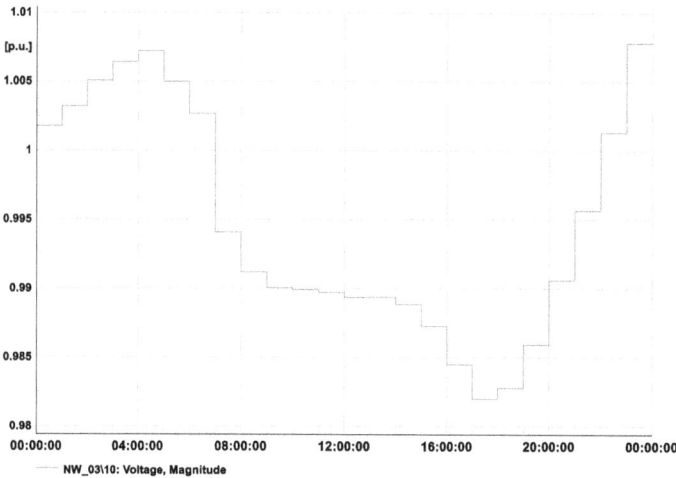

FIGURE 14.9
The voltage magnitude output information.

14.4 Tübingen_Germany 110kV Grid Example

There are 75 feeders in Tübingen, Germany's medium voltage network, which is extended from 110 kV to 410 V. This network, which you can upload from the book's end, has 950 loads, 1840 lines, and 5300 buses and terminals. See Figure 14.11. Finding the network's open loop points is a typical analysis in MV and LV distribution networks. In addition to lowering the short circuit level, an effort is made to open each loop in the network in order to lower network losses. Special algorithms are required to quickly identify these open points or closed switches because there are a lot of these candidate points.

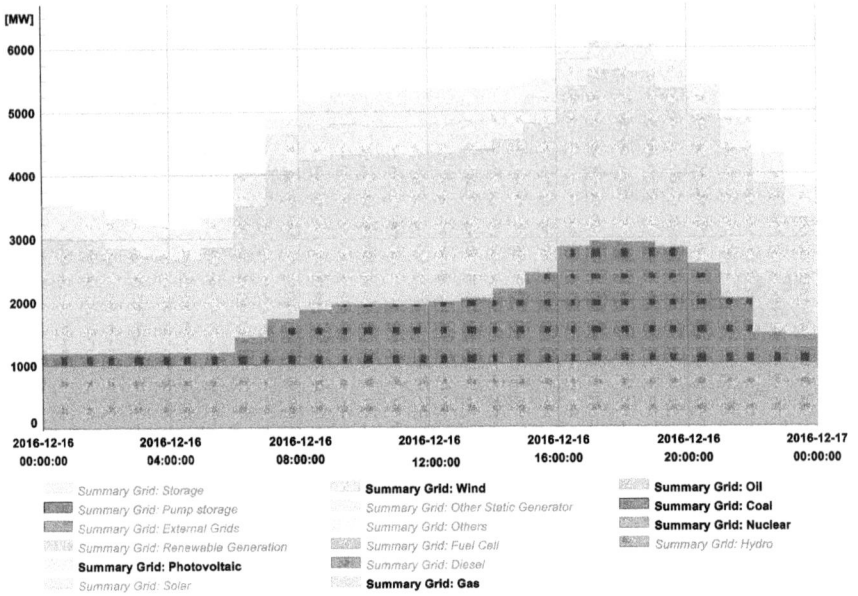

FIGURE 14.10
The power output information.

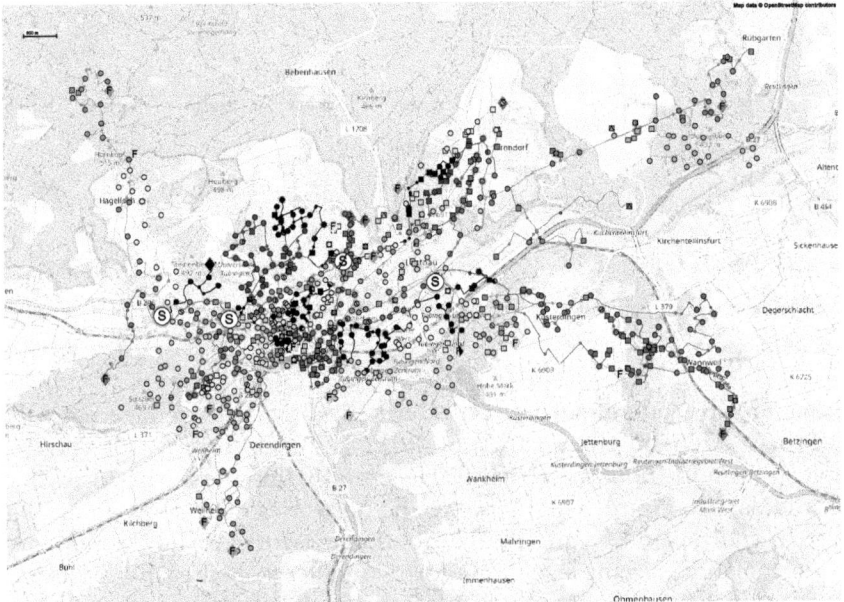

FIGURE 14.11
A medium-voltage distribution network.

TABLE 14.2

Necessary Switching Actions, From Initial to Optimized
Configuration)

Switch	Action
| SW_0804 [TRFSTAT_391]	OPEN
| SW_0951 [TRFSTAT_668]	OPEN
| SW_0995 [SUB_3]	OPEN
| SW_1181 [TRFSTAT_293]	OPEN
| SW_1585 [TRFSTAT_684]	OPEN
| SW_1816 [TRFSTAT_240]	OPEN
| SW_2197 [TRFSTAT_544]	OPEN
| SW_0807 [TRFSTAT_393]	CLOSE
| SW_0943 [TRFSTAT_650]	CLOSE
| SW_1424 [TRFSTAT_533]	CLOSE
| SW_1584 [TRFSTAT_684]	CLOSE
| SW_1820 [TRFSTAT_245]	CLOSE
| SW_2055 [TRFSTAT_431]	CLOSE
| SW_2198 [TRFSTAT_544]	CLOSE

The Tie Open Point Optimization software's output is shown in
Table 14.2. Table 14.3 displays the outcome of the opening and closing
of the specified open and closed switches. The benefits of using this
algorithm in the network include decreased losses and minimal voltage
fluctuations. This algorithm is used nearly daily by network operators.

14.5 400 MW Offshore Wind Farm Linked to HVDC Line

An HVDC line connects a 400 MW offshore wind farm to the main grid
in this simulation. As shown in Figures 14.12–14.14, a set of 5 MW wind
turbines in sets of 10 and 20 (Figure 14.12) are connected to a 150 kV, 1.4 kV,
100 km long HVDC line by two 240 MW three-winding transformers. The
transmitted power is then transferred to the grid by a 450 MW, 380/110 kV
transformer. In the event of a symmetrical three-phase short circuit on the
offshore wind farms' common bus, the output results are analyzed in two
simulation modes (RMS) and (EMT).

Thermal, control, and electromechanical device dynamics are taken
into account by the balanced RMS simulation approach. The passive electrical network is represented in a symmetric, steady-state manner. Only
the basic elements of voltages and currents are taken into account when
using this representation. The low level of detail makes it easier to analyze
the RMS simulation of the EMT.

In order to account for the dynamic behavior of passive network elements, the EMT simulation represents voltages and currents by their

TABLE 14.3

The Output of the Tie Open Point Optimization

Feeder	Losses [MW]	Total Load [MW]	Number of Customers [-]	Max. Voltage Drop [%]	Max. Voltage Rise [%]	Maximum Voltage [p.u.]
TOTAL				WORST CASE		
Before Optimization	1.7474477	172.938000	89233.000000	0.206844	0.066053	1.003847
After Optimization	1.7439213	172.938000	89233.000000	0.468356	0.110458	1.005360
Difference	-0.008263	0.000000	0.000000	0.261512	0.044405	0.001512
Difference [%]	-0.472879	0.000000	0.000000	126.429694	67.227257	0.150632

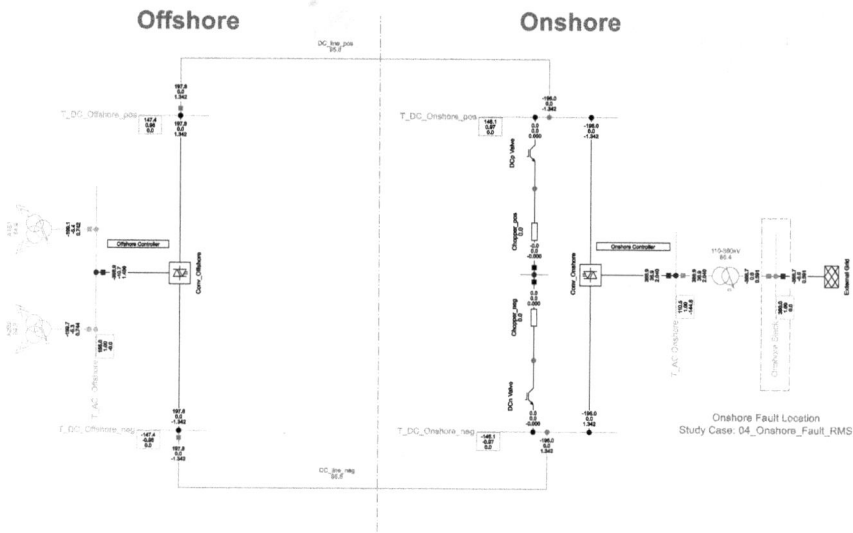

FIGURE 14.12
The first part of the network, the HVDC line connected to the external grid.

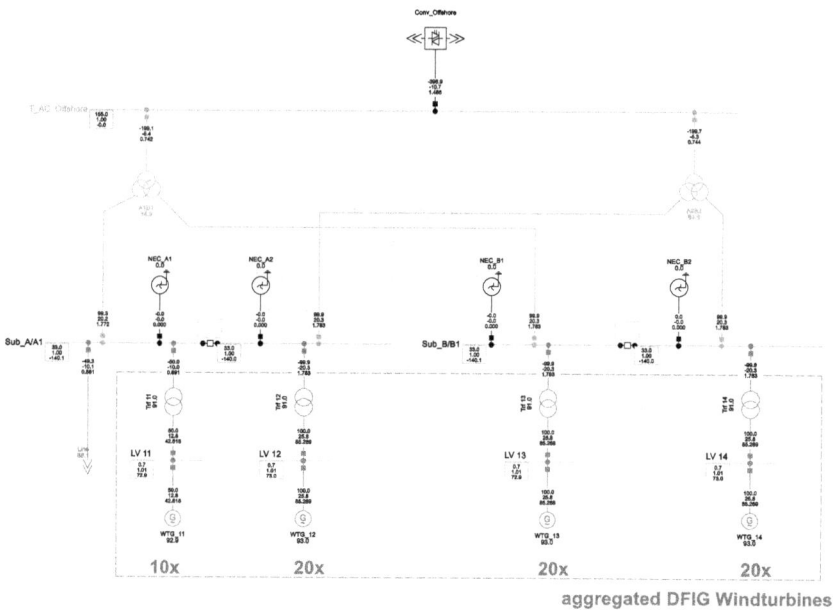

aggregated DFIG Windturbines

FIGURE 14.13
The second part of the network, the aggregated DFIG wind turbines.

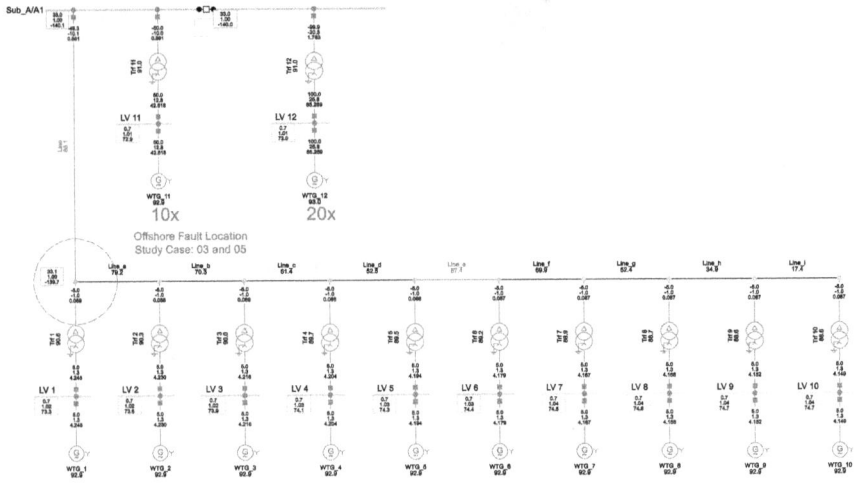

FIGURE 14.14
The third part of the network, the wind farm with details opened.

instantaneous values. All phases and all defined events (symmetrical and asymmetrical) can be simulated thanks to the high level of detail used to represent the modeled network. Longer-term transients can also be simulated using the EMT function. However, the calculation time increases because the integration step size must be much smaller than in the case of a steady-state representation because the passive network elements are represented dynamically.

In Figure 14.14, a symmetrical three-phase short circuit has occurred on the wind farm's common bus. The transmission line switch of this farm opens to the grid after 0.15 second. In order to plot the DC voltage, transmitted powers, and currents of the HVDC line, as well as the voltage and current of the wind farm connection line, two kinds of RMS-EMT simulations are run with this fault. See Figures 14.15–14.18. It is evident from comparing the two simulations that the EMT simulation is more detailed than the RMS. The quicker RMS simulation can provide the needed information in the simulation if the details are not crucial.

14.6 Khorasan_Iran Grid Example

This section presents a transient stability analysis of a section of the Khorasan province network. Four external grids connect this network to the rest of the Iranian network. There are 94 buses, 753 terminals, 47 lines, 2-w Trfs. 8, 3-w Trfs. 17, 6 syn. machines, and 27 loads. See Figure 14.19.

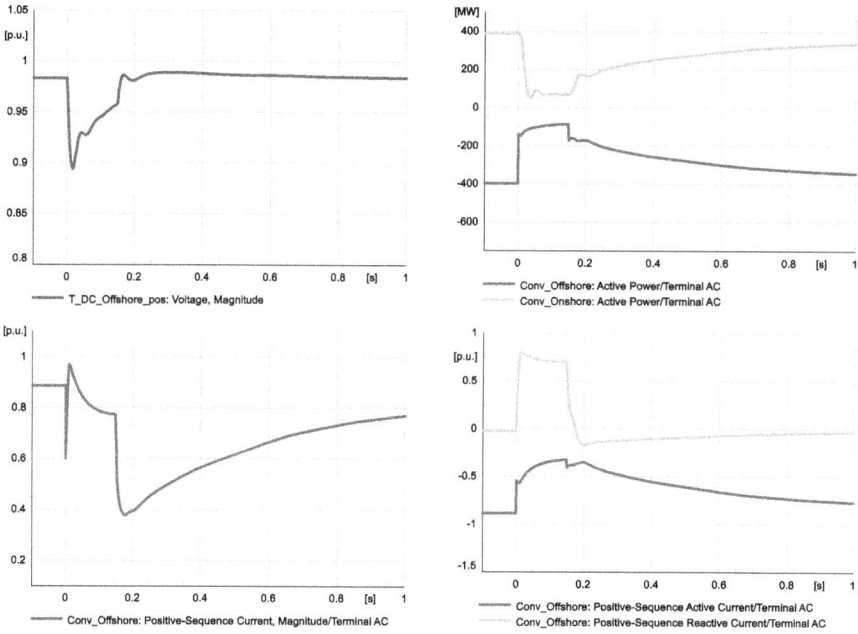

FIGURE 14.15
RMS Simulation, voltage, and current of HVDC line.

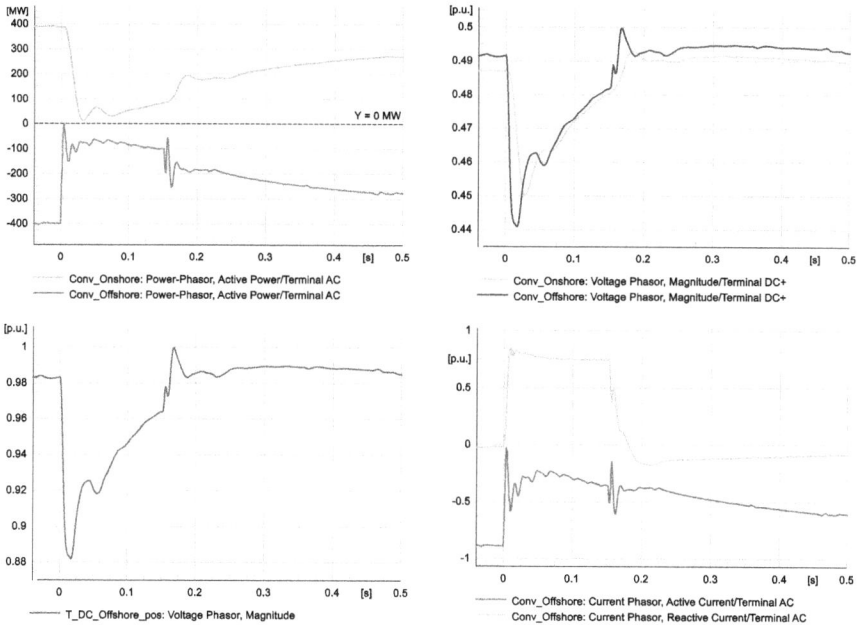

FIGURE 14.16
EMT Simulation, voltage, and current of HVDC line.

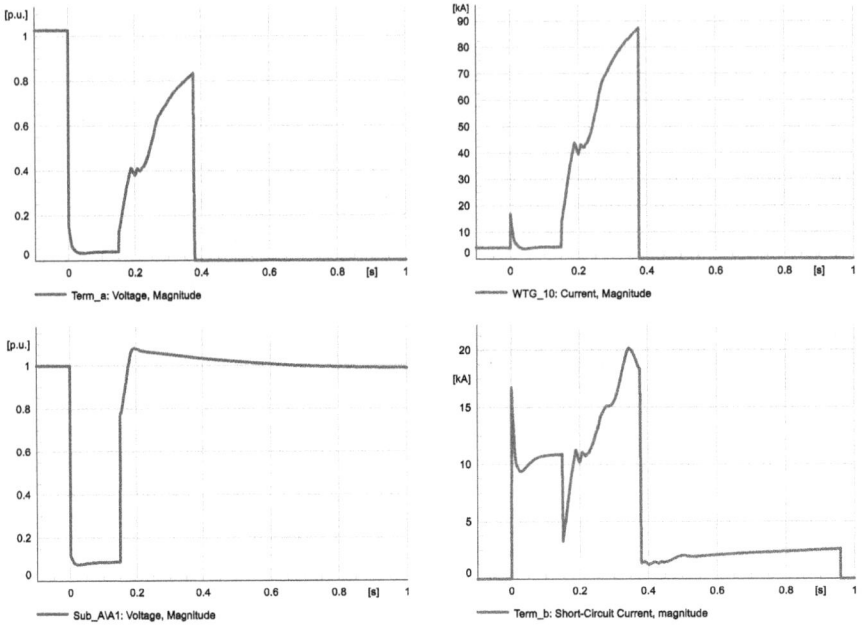

FIGURE 14.17
RMS Simulation, voltage, and current of a wind farm transmission line.

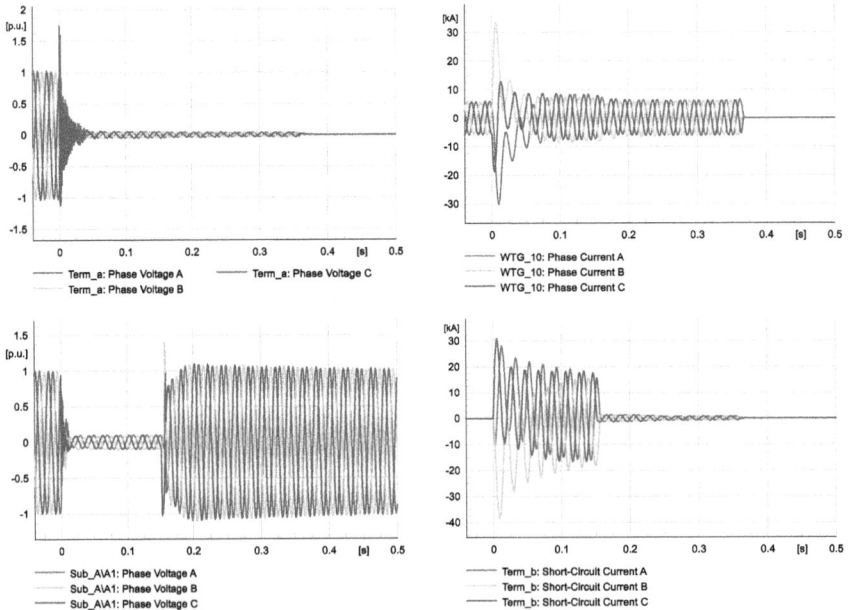

FIGURE 14.18
EMT Simulation, voltage and current of a wind farm transmission line.

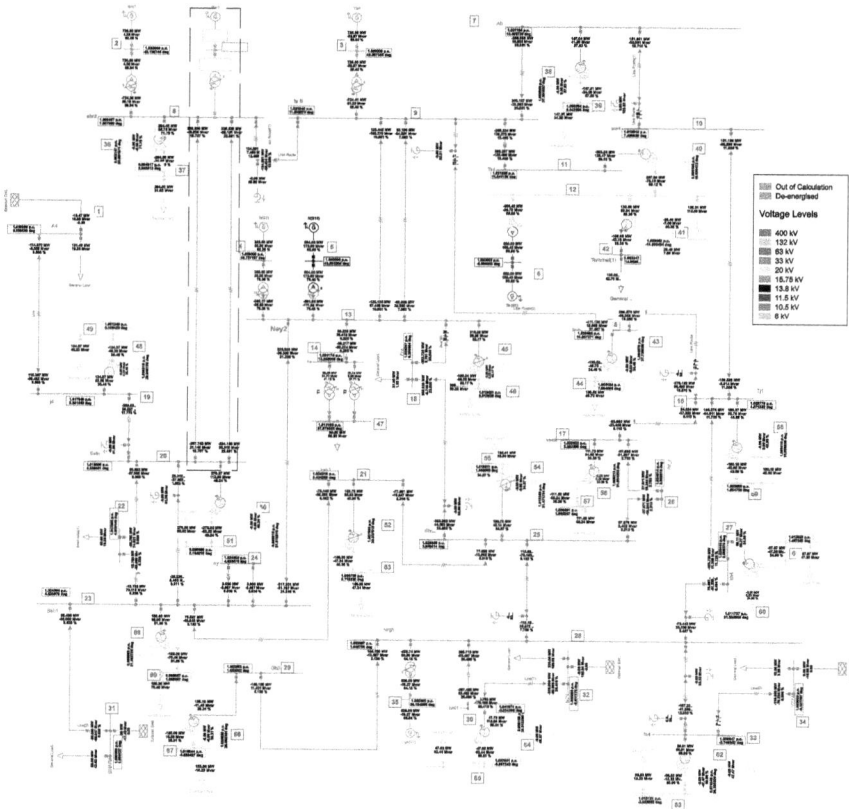

FIGURE 14.19
Khorasan provience_Iran Grid Example.

The network file is available for download at the book's conclusion. Here, one of the parallel transmission lines near Bus 8's generator (shown in Figure 14.19) has experienced a three-phase short circuit. The line switches are opened to clear the short circuit after 0.2 seconds. It is evident from Figure 14.20 that the network exhibits transient stability throughout the initial oscillation. In this instance, the circuit has been without power for one of the generators that is connected to bus 8. The generator inside the dashed line, which is connected to bus 8, enters the circuit in the second scenario.

Figure 14.21 plots the generator angles once more using the same short circuit and cut-off time. The network is momentarily unstable in this instance. The rise in the number of generators near the short circuit is the cause of this instability. The network cannot transfer the power absorbed during the fault and cannot return to a stable state once the fault

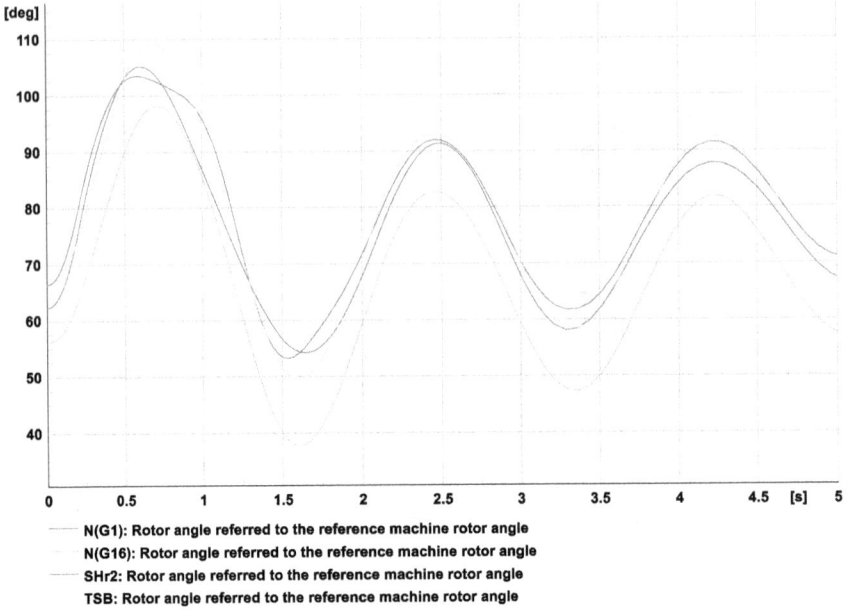

N(G1): Rotor angle referred to the reference machine rotor angle
N(G16): Rotor angle referred to the reference machine rotor angle
SHr2: Rotor angle referred to the reference machine rotor angle
TSB: Rotor angle referred to the reference machine rotor angle

FIGURE 14.20
Stable generator-less grid.

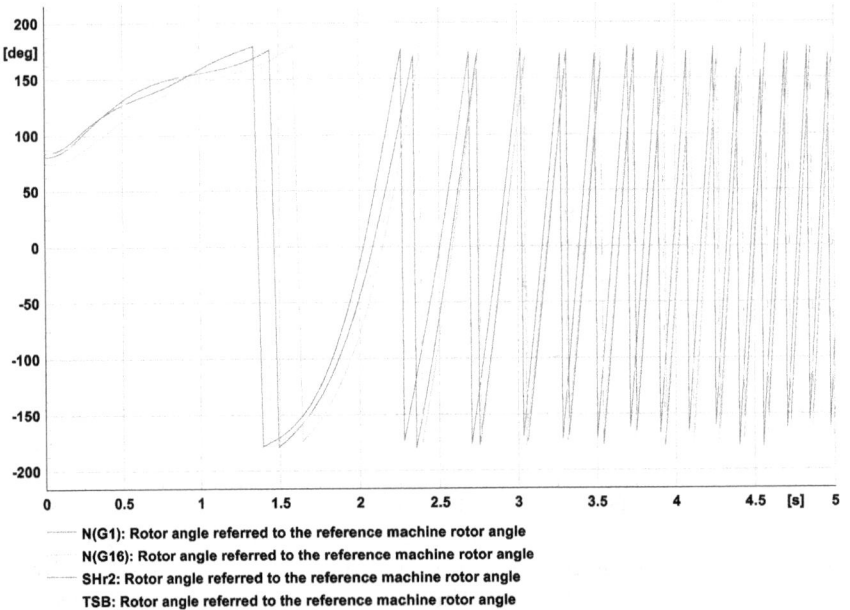

N(G1): Rotor angle referred to the reference machine rotor angle
N(G16): Rotor angle referred to the reference machine rotor angle
SHr2: Rotor angle referred to the reference machine rotor angle
TSB: Rotor angle referred to the reference machine rotor angle

FIGURE 14.21
Unstable generator-full grid.

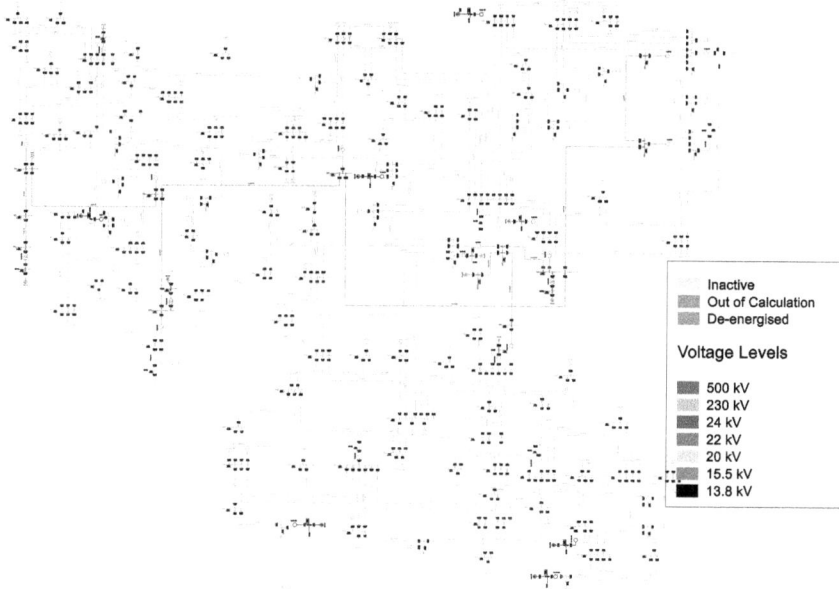

FIGURE 14.22
IEEE 118BUS_modified example.

is removed because of the enormous oscillation caused by the increase in the number of generators near the fault.

14.7 IEEE 118BUS_modified

In this chapter's next simulation, we look at the well-known IEEE 118-bus modified network, which has over 118 buses! 2022 terminals, 177 lines, 137 substations, 137 busbars, 28 two-winding transformers, 19 synchronous machines, 91 loads, 14 reactors, and shunt capacitors make up this network. The voltage levels in this network range from 500 to 13.8 kV, as shown in Figure 14.22. Four lines—11-13, 13-15, 14-15, and 12-14—losing simultaneously in the first second and then returning to the network in ten seconds is the scenario taken into consideration in this network.

It's interesting to note that, due to the lack of a controller, when lines are lost, the grid frequency and generator speed rise (Figure 14.23) and do not return to their initial values until the lost lines are restored to the grid within ten seconds. In this instance, it takes a few seconds for the grid to return to its initial nominal values.

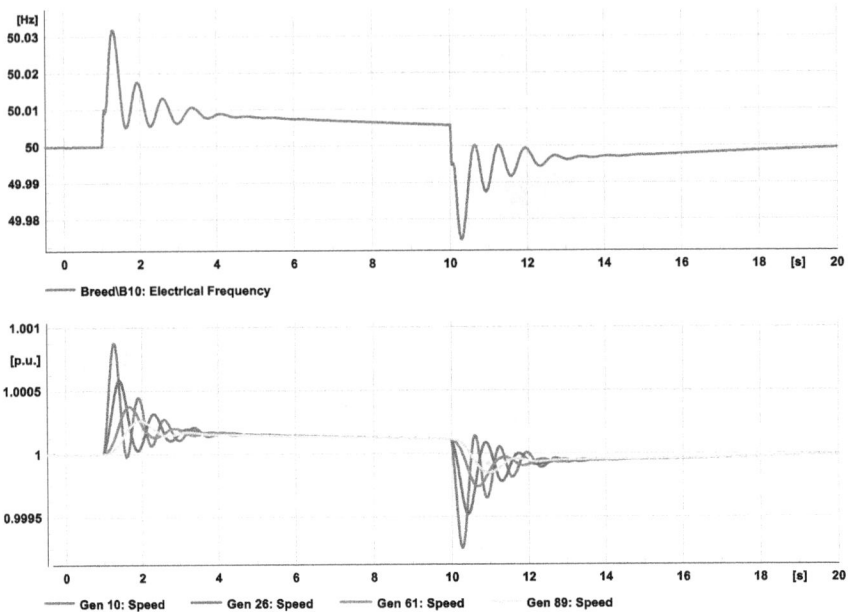

FIGURE 14.23
Frequency and generators speed at the entry and exit of the lines.

14.8 Extra Large Distribution Test Networks in Iran

The main distinction between these two networks and other networks is the sheer quantity of buses (9000–11,000) and the distribution-level realism of these networks. These two networks' simulation times are so lengthy that they can be used to test the software's ability to perform numerous calculations. The ability of algorithms to perform calculations is evident in large real networks, but when the number of network buses is small, the difference in algorithms in hundredths of a second is unclear.

Additionally, these test networks are very helpful to researchers in the power distribution field because they offer a standard topology with all the necessary real data. This allows them to apply various analyses, including load flow, reliability, voltage stability, and rearrangement, to the network, evaluate the results, and practically generalize them to other networks. These test networks satisfy every requirement listed above.

These networks are fed from various distribution substations and feature a large number of feeders with maneuvering points in between. These test networks also include: unbalanced and single-phase loads;

various protection devices such as autoreclosers, sectioners, fault detectors, and cut-out fuses; reliability data (exact number of subscribers on each load point, average number and time of outages of each feeder); a specific percentage of each of the following: domestic, commercial, industrial, agricultural, and public loads, as well as a certain percentage of each at each load point; and distributed photovoltaic production in feeders with load profiles and their production on a daily and seasonal basis.

Different kinds of balanced and unbalanced load flow programs can be implemented in these networks as time series or static loads. Additionally, the impact of loads on voltage is considered. These networks can benefit from the following: power quality improvement, reliability improvement, network protection, network automation, network expansion, adding switches, reducing imbalance, fault location, etc.; reliability analysis; network protection and voltage stability; implementation of various distribution network optimization algorithms with the aim of reducing losses, such as capacitor placement and distributed generation; network reorganization; voltage-wise control and load response; etc.

The files for these two distribution networks are available for download in the book's appendix. For instance, the "Tie Open Point Optimization" algorithm has been applied to this network. Thirty minutes later, the outcome was evident. Although the computation time is obviously highly computer-dependent, all of the earlier networks took less than a few seconds, with the exception of this one. In actuality, the calculation time in large networks grows exponentially with the number of buses and does not have a linear relationship with the number of buses. Table 14.4 displays a portion of the outcome.

TABLE 14.4

Tie Open Point Optimization Algorithm

Tie Open Point Optimization		Annex: / 1
Study Case: Study Case		

Necessary Switching Actions (From Initial to Optimized Configuration)

Name	Action
Switch/L1661	OPEN
Switch/L2998	OPEN
Switch/L4766	OPEN
Switch/L7692	OPEN
Switch/L3210	OPEN
Switch/L3722	OPEN
Switch/L4154	OPEN
Tie Switch/L7960	CLOSE
Tie Switch/L7961	CLOSE
Tie Switch/L7963	CLOSE
Tie Switch/L7964	CLOSE
Tie Switch/L7962	CLOSE
Tie Switch/L3196	CLOSE
Tie Switch/L4465	CLOSE

Feeders Results

(Continued)

TABLE 14.4 (*Continued*)

Tie Open Point Optimization Algorithm

Feeder	Losses [MW]	Total Load [MW]	Number of Customers [-]	Max.Voltage Drop [%]	Max. Voltage Rise [%]	Minimum Voltage [p.u.]	Maximum Voltage [p.u.]
Feeder1							
Before Optimization	1.949587	8.292207	8136.000000	25.671506	0.000000	0.680827	0.937287
After Optimization	0.924011	6.429538	4400.000000	17.785324	0.000000	0.779478	0.957154
Difference	-1.025576	-1.862669	-3736.000000	-7.886182	0.000000	0.098652	0.019867
Feeder2							
Before Optimization	0.898199	5.092422	3153.000000	22.815350	0.000000	0.723220	0.951303
After Optimization	0.937898	5.187737	4159.000000	23.043869	0.000000	0.706071	0.936436
Difference	0.039699	0.095316	1006.000000	0.228520	0.000000	-0.017150	-0.014867
Feeder3							
Before Optimization	0.455905	4.389819	3295.000000	14.129737	0.000000	0.810077	0.951317
After Optimization	0.795083	5.813108	3492.000000	16.797966	0.000000	0.768530	0.936431
Difference	0.339178	1.423289	197.000000	2.668229	0.000000	-0.041547	-0.014886

14.9 Summary

This chapter focused on using DIgSILENT PowerFactory software for simulating large electrical networks, including the extensive Texas grid and several other significant grids in Germany, Iran, and a modified IEEE network. It covered various analyses such as transient stability and frequency analysis, providing insights into each network's performance. Users can replicate the simulations by importing files from the book's appendix, with additional resources available for further study.

References

1. Eidiani, M., "A rapid state estimation method for calculating transmission capacity despite cyber security concerns," *IET Generation, Transmission and Distribution*, 2023, 17(20), pp. 4480–4488, https://doi.org/10.1049/gtd2.12747

2. Zeynal, H., Eidiani, M., Yazdanpanah, D., "Intelligent substation automation systems for robust operation of smart grids," *2014 IEEE Innovative Smart Grid Technologies—Asia* (ISGT ASIA), 2014, pp. 786–790, https://doi.org/10.1109/ISGT-Asia.2014.6873893

3. Eidiani, M., Shanechi, M.H.M., Vaahedi, E. "Fast and accurate method for computing FCTTC (first contingency total transfer capability)," *Proceedings. International Conference on Power System Technology*, 2002, vol. 2, pp. 1213–1217, https://doi.org/10.1109/ICPST.2002.1047595

4. Eidiani, M., Kargar, M., Zeynal, H., "Interactive use of D-STATCOM and storage resource to maintain microgrid stability for commercial systems," In: Sivaraman, P., Sharmeela, C., Sanjeevikumar, P. (eds) *Microgrids for Commercial Systems*, Wiley and Scrivener Publishing, 2024, pp. 241–270. https://doi.org/10.1002/9781394167319.ch10

5. Yazdi, S. H., Rouzbehi, K., Candela, J. I., Milimonfared, J., Rodriguez, P., "Flexible HVDC transmission systems small signal modelling: A case study on CIGRE Test MT-HVDC grid," *IECON 2017-43rd Annual Conference of the IEEE Industrial Electronics Society*, 2017.

6. Yazdi, S. H., Rouzbehi, K., Candela, J. I., Milimonfared, J., Rodriguez, P., "Analysis on impacts of the shunt conductances in multi-terminal HVDC grids optimal power-flow," *IECON 2017-43rd Annual Conference of the IEEE Industrial Electronics Society*, 2017.

7. Rouzbehi, K., Miranian, A., Luna, A., "Optimized control of multiterminal dc grids using particle swarm optimization," *EPE journal*, vol. 24, 2014.

Appendix: Download Link of Network Files

You can get in touch with the book's authors or download the files needed for each chapter from the link below.

(eidiani@yahoo.com).

DIgSILENT PowerFactory Files.zip

https://drive.google.com/file/d/1JZNsvjuAIMy0wesbEgBD999vA1sRi-q8/view?usp=sharing

Bibliography

Baydokhty, M.E., Eidiani, M., Zeynal, H., Torkamani, H., Mortazavi, H., "Efficient generator tripping approach with minimum generation curtailment based on fuzzy system rotor angle prediction," *Przeglad Elektrotechniczny*, 2012, 88(9 A), pp. 266–271.

Eidiani, M., "Assessment of voltage stability with new NRS," *2008 IEEE 2nd International Power and Energy Conference*, 2008, pp. 494–496, https://doi.org/10.1109/PECON.2008.4762525.

Eidiani, M., "A new method for assessment of voltage stability in transmission and distribution networks," *International Review of Electrical Engineering*, 2010, 5(1), pp. 234–240.

Eidiani, M., "A reliable and efficient method for assessing voltage stability in transmission and distribution networks," *International Journal of Electrical Power and Energy Systems*, 2011, 33(3), pp. 453–456, https://doi.org/10.1016/j.ijepes.2010.10.007.

Eidiani, M., "An efficient differential equation load flow method to assess dynamic available transfer capability with wind farms," *IET Renewable Power Generation*, 2021a, 15, pp. 3843–3855, https://doi.org/10.1049/rpg2.12299.

Eidiani, M., "A reliable and efficient holomorphic approach to evaluate dynamic available transfer capability," *International Transactions on Electrical Energy Systems*, 2021b, 31(11), p. e13031, https://doi.org/10.1002/2050-7038.13031.

Eidiani, M., "A new load flow method to assess the static available transfer capability," *Journal of Electrical Engineering and Technology*, 2022a, 17(5), pp. 2693–2701, https://doi.org/10.1007/s42835-022-01105-3.

Eidiani, M., "Atc evaluation by CTSA and POMP, two new methods for direct analysis of transient stability," *IEEE/PES Transmission and Distribution Conference and Exhibition*, 2002b, vol. 3, pp. 1524–1529, https://doi.org/10.1109/TDC.2002.1176824.

Eidiani, M., "Applying optimization techniques to develop a renewable energy supply map," In: Fathi, M., Zio, E., Pardalos, P.M. (eds) *Handbook of Smart Energy Systems*. Springer, Cham, 2022c, https://doi.org/10.1007/978-3-030-72322-4_61-1.

Eidiani, M., "Integration of renewable energy sources," In: Fathi, M., Zio, E., Pardalos, P.M. (eds) *Handbook of Smart Energy Systems*. Springer, Cham, 2022d, https://doi.org/10.1007/978-3-030-72322-4_41-1.

Eidiani, M., "A new hybrid method to assess available transfer capability in AC–DC networks using the wind power plant interconnection," *IEEE Systems Journal*, 2023a, 17(1), pp. 1375–1382, https://doi.org/10.1109/JSYST.2022.3181099.

Eidiani, M., "A rapid state estimation method for calculating transmission capacity despite cyber security concerns," *IET Generation, Transmission and Distribution*, 2023b, 17(20), pp. 4480–4488, https://doi.org/10.1049/gtd2.12747.

Eidiani, M., "Modeling renewable energy resources using DIgSILENT PowerFactory software," In: Chenniappan, S., Padmanaban, S., Palanisamy, S. (eds) *Power Systems Operation with 100% Renewable Energy Sources*, 2024a, pp. 165–202. https://doi.org/10.1016/B978-0-443-15578-9.00013-3

Eidiani, M., "Online dynamic ATC computation with large-scale wind farms," *Electrical Engineering*, 2024b, 106, pp. 5677–5684, https://doi.org/10.1007/s00202-024-02325-8.

Eidiani, M., "The effect of power system strength on the calculation of available transmission capacity," In: *Power System Strength: Evaluation Methods, Best Practice, Case Studies, and Applications*, 2024c, pp. 137–174, https://doi.org/10.1049/PBPO247E_ch7.

Eidiani, M., Asadi, S.M., Faroji, S.A., Velayati, M.H., Yazdanpanah, D., "Minimum distance, a quick and simple method of determining the static ATC," *2008 IEEE 2nd International Power and Energy Conference*, 2008, pp. 490–493, https://doi.org/10.1109/PECON.2008.4762524.

Eidiani, M., Ashkhane, Y., Khederzadeh, M., "Reactive power compensation in order to improve static voltage stability in a network with wind generation," *2009 International Conference on Computer and Electrical Engineering, ICCEE 2009*, 2009, 1, pp. 47–50, p. 5380672, https://doi.org/10.1109/ICCEE.2009.239.

Eidiani, M., Badokhty, M.E., Ghamat, M., Zeynal, H., "Improving transient stability using combined generator tripping and braking resistor approach," *International Review on Modelling and Simulations*, 2011a, 4(4), pp. 1690–1699.

Eidiani, M., Baydokhty, M.E., Ghamat, M., Zeynal, H., Mortazavi, H., "Transient stability enhancement via hybrid technical approach," *2011 IEEE Student Conference on Research and Development*, 2011b, pp. 375–380, https://doi.org/10.1109/SCOReD.2011.6148768

Eidiani, M., Buygi, M.O., Ahmadi, S., "CTV, complex transient and voltage stability: A new method for computing dynamic ATC," *International Journal of Power and Energy Systems*, 2006, 26(3), pp. 296–304, https://doi.org/10.2316/Journal.203.2006.3.203-3597.

Eidiani, M., Ghavami, A., "New network design for simultaneous use of electric vehicles, photovoltaic generators, wind farms and energy storage," *2022 9th Iranian Conference on Renewable Energy & Distributed Generation (ICREDG)*, pp. 1–5, https://doi.org/10.1109/ICREDG54199.2022.9804534.

Eidiani, M., Ghavami, A., Zeynal, H., Zakaria, Z., "Performance of Monofacial and Bifacial Solar System for 5MW grid-connected solar power plant with fixed tilt and tracking: Design, simulation and investigating the effect of temperature," *2024 28th International Electrical Power Distribution Conference, EPDC 2024*, 2024a.

Eidiani, M., Ghavami, A., Zeynal, H., Zakaria, Z., "Performance of monofacial and bifacial solar system for 5MW grid-connected solar power plant with fixed tilt and tracking: Design, simulation and investigating the effect of temperature," *2024 28th International Electrical Power Distribution Conference (EPDC)*, Zanjan, Iran, Islamic Republic of, 2024b, pp. 1–7, doi: 10.1109/EPDC62178.2024.10571696.

Eidiani, M., Heidari, V., *Fundamentals of Power Systems Analysis 1: Problems and Solutions*, Taylor & Francis Group, CRC Press, 2023, pp. 1–215, https://doi.org/10.1201/9781003394433.

Eidiani, M., Kargar, M., "Frequency and voltage stability of the microgrid with the penetration of renewable sources," *2022 9th Iranian Conference on Renewable Energy & Distributed Generation (ICREDG)*, pp. 1–6, https://doi.org/10.1109/ICREDG54199.2022.9804542.

Eidiani, M., Mahnani, S., "Optimally and independent planned microgrid with solar-wind and biogas hybrid renewable systems by HOMER," *Majlesi Journal of Electrical Engineering*, 2023, 17(2), pp. 153–158, https://doi.org/10.30486/mjee.2023.1974843.1021.

Eidiani, M., Rouzbehi, K., *Advanced Topics in Power Systems Analysis: Problems, Methods, and Solutions*, Taylor & Francis Group, CRC Press, 2024, pp. 1–120.

Eidiani, M., Rouzbehi, K., *Fundamentals of Power System Transformers Modeling, Analysis, and Operation*, Taylor & Francis Group, CRC Press, 2025, pp. 1–127

Eidiani, M., Shahdehi, N.A., Zeynal, H., "Improving dynamic response of wind turbine driven DFIG with novel approach," *2011 IEEE Student Conference on Research and Development*, 2011, pp. 386–390, https://doi.org/10.1109/SCOReD.2011.6148770.

Eidiani, M., Shanechi, M.H.M., "FAD-ATC: A new method for computing dynamic ATC," *International Journal of Electrical Power and Energy Systems*, 2006, 28(2), pp. 109–118, https://doi.org/10.1016/j.ijepes.2005.11.004.

Eidiani, M., Shanechi, M.H.M., Vaahedi, E., "Fast and accurate method for computing FCTTC (first contingency total transfer capability)," *Proceedings. International Conference on Power System Technology*, 2002, vol. 2, pp. 1213–1217 https://doi.org/10.1109/ICPST.2002.1047595.

Eidiani, M., Yazdanpanah, D., "Minimum distance, a quick and simple method of determining the static ATC," *Journal of Electrical Engineering*, 2011, 11(2), pp. 95–101.

Eidiani, M., Zeynal, H., "New approach using structure-based modeling for the simulation of real power/frequency dynamics in deregulated power systems," *Turkish Journal of Electrical Engineering and Computer Sciences*, 2014, 22(5), pp. 1130–1146, https://doi.org/10.3906/elk-1208-90

Eidiani, M., Zeynal, H., "A fast holomorphic method to evaluate available transmission capacity with large scale wind turbines," *9th Iranian Conference on Renewable Energy & Distributed Generation (ICREDG)*, 2022a, pp. 1–5, https://doi.org/10.1109/ICREDG54199.2022.9804527.

Eidiani, M., Zeynal, H., "Determination of online DATC with uncertainty and state estimation," *2022 9th Iranian Conference on Renewable Energy & Distributed Generation (ICREDG)*, 2022b, pp. 1–6, https://doi.org/10.1109/ICREDG54199.2022.9804581.

Eidiani, M., Zeynal, H., "An effective method to determine the available transmission capacity with variable frequency transformer," *International Transactions on Electrical Energy Systems*, 2023, 2023, p. 8404284, https://doi.org/10.1155/2023/8404284.

Eidiani, M., Zeynal, H., Ghavami, A., Zakaria, Z., "Comparative analysis of mono-facial and bifacial photovoltaic modules for practical grid-connected solar power plant using PVsyst," *2022 IEEE International Conference on Power and Energy (PECon)*, 2022, pp. 499–504, https://doi.org/10.1109/PECon54459.2022.9988872.

Eidiani, M., Zeynal, H., Ghavami, A., Zakaria, Z., "Performance of Bifacial Solar System for 1MW grid-connected solar power plant in Mashhad, Iran: Design, Simulation and Economic Analysis," *2024 9th International Conference on Technology and Energy Management, ICTEM 2024*, 2024.

Eidiani, M., Zeynal, H., Shaaban, M., "A detailed study on prevailing ATC methods for optimal solution development," *2022 IEEE International Conference on Power and Energy (PECon)*, 2022, pp. 299–303, https://doi.org/10.1109/PECon54459.2022.9988775.

Eidiani, M., Zeynal, H., Zadeh, A.K., Mansoorzadeh, S., Nor, K.M., "Voltage stability assessment: An approach with expanded Newton Raphson-Sydel," *2011 5th International Power Engineering and Optimization Conference*, 2011, pp. 31–35, https://doi.org/10.1109/PEOCO.2011.5970424.

Eidiani, M., Zeynal, H., Zadeh, A.K., Nor, K.M., "Exact and efficient approach in static assessment of Available Transfer Capability (ATC)," *2010 IEEE International Conference on Power and Energy*, 2010, pp. 189–194, https://doi.org/10.1109/PECON.2010.5697580.

Eidiani, M., Zeynal, H., Zakaria, Z., "An efficient holomorphic based available transfer capability solution in presence of large scale wind farms," *2022 IEEE International Conference in Power Engineering Application (ICPEA)*, 2022a, pp. 1–5, https://doi.org/10.1109/ICPEA53519.2022.9744711.

Eidiani, M., Zeynal, H., Zakaria, Z., "Development of online dynamic ATC calculation integrating state estimation," *2022 IEEE International Conference in Power Engineering Application (ICPEA)*, 2022b, pp. 1–5, https://doi.org/10.1109/ICPEA53519.2022.9744694.

Eidiani, M., Zeynal, H., Zakaria, Z., "A comprehensive study on the renewable energy integration using DIgSILENT," *2023 IEEE 3rd International Conference in Power Engineering Applications (ICPEA)*, Putrajaya, Malaysia, 2023a, pp. 197–201, https://doi.org/10.1109/ICPEA56918.2023.10093153.

Eidiani, M., Zeynal, H., Zakaria, Z., "An efficient method for available transfer capability calculation considering cyber-attacks in power systems," *2023 IEEE 3rd International Conference in Power Engineering Applications: Shaping Sustainability Through Power Engineering Innovation, ICPEA 2023*, 2023b, pp. 127–130, https://doi.org/10.1109/ICPEA56918.2023.10093168.

Eidiani, M., Zeynal, H., Zakaria, Z., Shaaban, M., "Analysis of optimization methods applied for renewable energy integration," *2023 IEEE 3rd International Conference in Power Engineering Applications (ICPEA)*, 2023, p. 1570861924.

Ghardashi, G., Gandomkar, M., Majidi, S., Eidiani, M., Dadfar, S., "Accuracy and speed improvement of microgrid Islanding detection based on PV using frequency-reactive power feedback method," *2022 International Conference on Protection and Automation of Power Systems (IPAPS)*, 2022, pp. 1–8, https://doi.org/10.1109/IPAPS55380.2022.9763190.

Momen, S., Hekmati, A., Majidi, S., Zand, Z., Zand, M., Nikoukar, J., Eidiani, M., "Energy harvesting for smart energy systems," In: Fathi, M., Zio, E., Pardalos, P.M. (eds) *Handbook of Smart Energy Systems*. Springer, Cham, 2022, https://doi.org/10.1007/978-3-030-72322-4_12-1.

Parhamfar, M., Eidiani, M., Abtahi, M., "Distributed energy storage system: Case study," In: Palanisamy, S., Chenniappan, S., Padmanaban, S. (eds) *Distributed Energy Storage Systems for Digital Power Systems*, Elsevier, 2025, pp. 395–422. https://doi.org/10.1016/B978-0-443-22013-5.00013-7

Zand, Z., Ghahri, M.R., Majidi, S., Eidiani, M., Nasab, M.A., Zand, M., "Smart grid and resilience," In: Fathi, M., Zio, E., Pardalos, P.M. (eds) *Handbook of Smart Energy Systems.* Springer, Cham, 2022, https://doi.org/10.1007/978-3-030-72322-4_178-1.

Zeynal, H., Eidiani, M., "Hydrothermal scheduling flexibility enhancement with pumped-storage units," *2014 22nd Iranian Conference on Electrical Engineering (ICEE)*, 2014, pp. 820–825, https://doi.org/10.1109/IranianCEE.2014.6999649.

Zeynal, H., Eidiani, M., Yazdanpanah, D., "Intelligent substation automation systems for robust operation of smart grids," *2014 IEEE Innovative Smart Grid Technologies - Asia (ISGT ASIA)*, 2014, pp. 786–790, https://doi.org/10.1109/ISGT-Asia.2014.6873893.

Zeynal, H., Hui, L.X., Jiazhen, Y., Eidiani, M., Azzopardi, B., "Improving Lagrangian Relaxation unit commitment with Cuckoo Search Algorithm," *2014 IEEE International Conference on Power and Energy (PECon)*, 2014, pp. 77–82, https://doi.org/10.1109/PECON.2014.7062417.

Zeynal, H., Jiazhen, Y., Azzopardi, B., Eidiani, M., "Flexible economic load dispatch integrating electric vehicles," *2014 IEEE 8th International Power Engineering and Optimization Conference (PEOCO2014)*, 2014a, pp. 520–525, https://doi.org/10.1109/PEOCO.2014.6814484.

Zeynal, H., Jiazhen, Y., Azzopardi, B., Eidiani, M., "Impact of electric vehicle's integration into the economic VAr dispatch algorithm," *2014 IEEE Innovative Smart Grid Technologies—Asia (ISGT ASIA)*, 2014b, pp. 780–785, https://doi.org/10.1109/ISGT-Asia.2014.6873892.

Zeynal, H., Zadeh, A.K., Nor, K.M., Eidiani, M., "Locational marginal price (LMP) assessment using hybrid active and reactive cost minimization," *International Review of Electrical Engineering*, 2010, 5(5), pp. 2413–2418.

Index

For Product Safety Concerns and Information please contact our EU
representative GPSR@taylorandfrancis.com
Taylor & Francis Verlag GmbH, Kaufingerstraße 24, 80331 München, Germany